AMSCO'S Science

Grade 7

Paul S. Cohen

Former Assistant Principal, Science
Franklin Delano Roosevelt High School
Brooklyn, New York

Anthony V. Sorrentino, D. Ed.

Former Director of Computer Services
Former Earth Science Teacher
Monroe-Woodbury Central School District
Central Valley, New York

Amsco School Publications, Inc.
315 Hudson Street / New York, NY 10013

The publisher wishes to acknowledge the helpful contributions of Melissa and Tom McFeely and Martin Solomon, who acted as consultants in the preparation of this book. In addition, we wish to thank Jules J. Weisler for his valuable contributions.

Text Design: Howard Petlack, A Good Thing, Inc.
Cover Design: Meghan Shupe
Editor: Madalyn Stone
Composition: Publishing Synthesis, Ltd., New York
Artwork: Hadel Studio

Please visit our Web site at: *www.amscopub.com*

When ordering this book, please specify:
R 129 H or Amsco's Science Grade 7, *Hardbound*

ISBN-978-1-56765-916-0 (Hardbound Edition)

Contents

Unit

2

Interactions Between Matter and Energy

Unit

3

Dynamic Equilibrium: The Human Animal

Unit

4

Dynamic Equilibrium: Other Organisms

To the Teacher

Amsco's Science: Grade 7, the second of a new, 3-book series for middle school science from Amsco School Publications, Inc., provides complete, clear, and concise coverage of the concepts taught in a yearlong 7th grade science class. As a primary text, *Amsco's Science: Grade 7* contains ample activities and exercises that will encourage the inquiry processes of science education. As a supplementary text, it can be used to reinforce conceptual understanding that more activity-based texts may lack.

The text consists of thirteen chapters divided into four units. Unit 1, Geology, describes how scientists gather and interpret evidence that Earth is continually changing. Unit 2, Interactions Between Matter and Energy, shows how the properties and interactions of matter and energy explain physical and chemical changes. In Unit 3, Dynamic Equilibrium: The Human Animal, the authors explain how human body systems function to maintain homeostasis. And in the last unit, Unit 4, Dynamic Equilibrium: Other Organisms, we learn how homeostasis is maintained in other organisms.

The text presents the major ideas of each topic in a clear manner that is easy to understand. It is aligned and corresponds with the National Standards (as well as those of New York City and New York State), and it can be used with any integrated science middle school curriculum. Each chapter is divided into lessons that allow teachers flexibility. You can teach the lesson in one day, or more, depending on the needs of your students. Conceptual understanding of topics is emphasized, and a balance of activities is presented to reinforce the topics.

Each chapter has an introductory page listing the contents of the chapter, a statement about what is in the chapter, a related career-planning feature, and a list of some Internet sites that provide additional knowledge or activities. Lesson titles are written in the form of a question and have clearly stated objectives. At the beginning of the lesson, a list of important vocabulary terms is presented—with a phonetic pronunciation guide—in the order in which they appear in the text. Each term is printed in **bold-faced** type in the lesson, and all of these terms are in the Glossary at the back of the book. Terms of lesser importance, or that have been introduced earlier, are in *italics*. Lessons consist of several topics that are visually explained with diagrams and pictures, as well as summarized in tables.

Each lesson is followed by Multiple Choice and Thinking and Analyzing Questions that reinforce the main points of the lesson. Many of the questions are taken directly from past exams. Activities and Skill Exercises related to the lesson provide additional reinforcement and an opportunity for students to do science at home or in school. Many of the activities are designed using easy-to-obtain materials commonly found around the home. Others require access to the Internet. Most lessons also contain a special boxed feature called "Interesting Facts About. . ." that presents little-known facts about the topic being discussed, or a related topic.

Finally, at the end of every chapter there are Review Questions. The questions are divided into four sections: Term Identification, Multiple Choice, Thinking and Analyzing, and a Crossword Puzzle. In the Term Identification questions, students are asked to match a word with its definition, and also to contrast it with another term. The Multiple Choice and Thinking and Analyzing questions are similar to published exam questions. The Thinking and Analyzing questions include short answer, diagram analysis, and reading comprehension questions. The Crossword Puzzle is a fun way of reviewing the important vocabulary of the chapter.

Unit 1
Geology

Part I—Essential Question

This unit focuses on the following essential question:

How do we as scientists gather and interpret evidence that Earth is continually changing?

Most processes that change Earth's surface occur without us realizing Earth is changing. For example, winds and streams move loose rock particles. Generally, these processes carry rock particles from higher locations to lower locations, tending to wear down the land to a flat surface. Other changes in the land occur more dramatically. Volcanic eruptions and earthquakes cause the land to rise and form mountains. The processes wearing down the surface and the processes building up the surface continually change Earth.

Scientists carefully study the processes that change Earth's surface. Most processes occur relatively slowly and cannot be seen over a human lifetime. However, over millions of years, mountains can be pushed up and even be worn back down to sea level. In this unit, you will learn how scientists gather and interpret evidence revealing how our planet is changing.

Scientists observe and measure geologic events and conduct experiments to gather information about Earth. They report their findings to the scientific community, thereby combining many small bits of knowledge into a better understanding of the past, present, and possibly even the future of Earth. Specifically, much evidence lies in the rocks, fossils, and landforms around us. Rocks and fossils contain clues of past geologic and biological events. They also reveal information about the environment associated with the events. Geologic and biological events can be put in order of occurrence, and a timeline of the events can be developed. Landforms also provide evidence of a changing Earth. Mountains, plains, and plateaus indicate the type of forces that changed the land in the past.

Part II—Chapter Overview

In Unit 1, you will learn how scientists gather information from rocks, fossils, and landforms. The information eventually becomes evidence that describes how Earth is continually changing, and what processes cause the changes. The evidence also led to a theory explaining the cause of earthquakes, volcanoes, and mountain-building processes.

Chapter 1 develops the concept of Earth's structure. You will learn that Earth consists of several layers, and that the layers differ in composition, temperature, and density. Surrounding Earth there are four zones referred to as spheres. Three of the spheres, the atmosphere (air), hydrosphere (water), and lithosphere (rock) constantly exchange energy and matter. The fourth sphere, the biosphere (life), relies upon interaction with the other three spheres for its existence.

Chapter 2 describes the composition and properties of minerals. It also explains how rocks are classified by their method of formation. You will learn about the processes that cause rocks to wear down and be carried away. A continuous cycle of the forming of rocks, the wearing down of those rocks, and the eventual reforming of those rocks again is described as the rock cycle.

Chapter 3 focuses on fossils and Earth's history. You will see how fossils provide evidence that life on Earth has changed from simple forms to complex forms over long periods of time. Fossils are also used to determine the relative age of rocks. The true age of rocks is determined using a different method. All this information helped produce a timeline of Earth's geologic and biological history called the geologic timetable.

Chapter 4 discusses the development of a theory unifying many geologic events. Scientists know Earth's surface is broken into pieces, or plates, and the plates move because heat is released from Earth's interior. Most earthquakes, volcanoes, and mountain-building processes occur along the boundaries of moving plates. Also in Chapter 4 is a lesson on topographic maps, which discusses how the shape and form of the land can be shown on a map.

Chapter 1

Earth as a System

Contents

A picture of Earth taken from space showing the atmosphere, hydrosphere, and lithosphere.

What Is This Chapter About?

Earth is a system. This means that our planet, Earth, consists of a collection of interworking parts. The parts are called spheres. The atmosphere, hydrosphere, and lithosphere consist of nonliving matter. The biosphere is made up of living things, and relies on the other spheres for its existence.

In this chapter you will learn:

1. The structure of Earth consists of three distinct layers: the crust, mantle, and core.

2. The atmosphere, hydrosphere, and lithosphere are three spheres that surround Earth.

3. The biosphere, or the sphere that contains life, interacts with the other three spheres.

Science in Everyday Life

Look around you. In what ways do humans use air, water, and rocks? We breathe in (inhale) oxygen from air and use water to grow food and clean our bodies. We use rocks to produce building materials such as stones and metals. Air, water, and rocks have many more uses. How many can you list?

Internet Sites:

http://www.enchantedlearning.com/subjects/astronomy/planets/earth/Inside.shtml The Enchanted Learning site provides information about the interior of Earth. You may also take tests and use worksheets to help you learn more about Earth.

http://www.moorlandschool.co.uk/earth/earths_structure.htm Dig deeper into Earth's structure. Find out how scientists know what's inside Earth. Take an Earth structure test.

http://www.geography4kids.com/ Geography 4 Kids covers many science topics, including Earth's structure, atmosphere, hydrosphere, and biosphere. This site also provides activities and quizzes to help you improve your science knowledge.

1.1

What Is the Structure of Earth?

Objectives
Describe how Earth formed.

Describe Earth's structure.

Terms
crust: Earth's thin outermost layer that is composed of solid rock

mantle: a layer of Earth that lies between the crust and the core

core: the center of Earth

Formation of Earth

About 4.5 billion years ago, dust and gas molecules came together in our area of the universe. Slowly, a rotating cloud-like region formed. Gravity caused the cloud to collapse, producing a star in the center and planets, satellites, comets, and asteroids revolving around the star. The star became our sun and one of the revolving planets became Earth. (Pictures and a short movie of this process can be seen at *http://cougar. jpl.nasa.gov/HR4796/anim.html.*)

The force of gravity pulling particles together and the decay of radioactive elements generated heat in the center of this newly developed planet. Early Earth was very hot and probably in a molten (melted, liquid-like) state similar to volcanic lava. As Earth continued to cool, dense elements, such as iron and nickel, settled to form a central core region. Less dense elements, such as silica and aluminum, moved to the surface. Density differences among these elements produced the different layers of Earth's structure. Today, after 4.5 billion years, heat that is still rising from inside Earth indicates that our planet is still cooling.

Internal Structure of Earth

Planet Earth has a layered internal structure. (See Figure 1.1-1.) Our knowledge of Earth's structure continuously improves. Geologists (scientists who study Earth) are most familiar with the outer layer of Earth because it is the most easy to study. Geologists learn about Earth's other layers by gathering information from earthquake vibrations, volcanic eruptions, calculations related to the force of gravity, and laboratory experiments.

Earth's outermost layer is the **crust**. The crust, composed mainly of solid igneous rock, makes up about 2 percent of Earth's

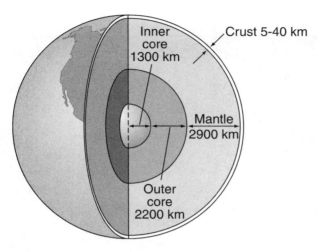

Figure 1.1-1. Cutaway section of Earth showing its internal structure.

volume. Compared to the other layers, the crust is very thin and brittle. In areas of Earth where the crust breaks, earthquakes occur. Under the oceans, the crust is about 5 km (3 miles) thick and consists of mostly dark-colored rocks rich in iron and

magnesium. Thicker crust lies under the continents. The average thickness of the continental crust is about 40 km (25 miles). Under the continents, the crust consists mostly of light-colored rocks rich in silica and aluminum.

Below the crust is a layer called the **mantle**. The mantle is about 2900 km (1800 miles) thick. The mantle makes up about 88 percent of Earth's volume. The area just below the crust is the upper mantle. The lower portion of the mantle is much thicker and consists of dense iron- and magnesium-rich rock material. Although the mantle is solid, internal heat and pressure allow it to flow like melting plastic. The mantle flows in large convection cells at the rate of about 2 cm (less than one inch!) per year. (See Figure 1.1-2.)

At Earth's center is the **core**, which contains both an *outer core* and an *inner core*. The core makes up about 10 percent of Earth's volume. The outer core is about 2200 km (1400 miles) thick and is thought to be

Figure 1.1-2. A slice of Earth showing its layered structure and the temperature and density range of each layer.

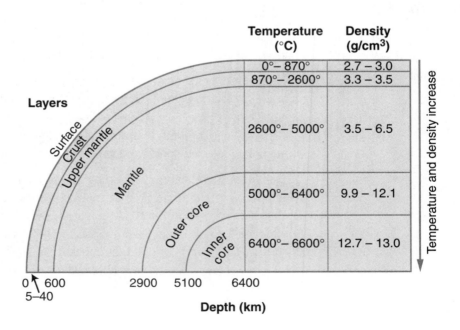

Temperature (°C)	Density (g/cm³)
0°– 870°	2.7 – 3.0
870°– 2600°	3.3 – 3.5
2600°– 5000°	3.5 – 6.5
5000°– 6400°	9.9 – 12.1
6400°– 6600°	12.7 – 13.0

Table 1.1-1. Structure of Earth

Layer	Thickness (km)	Temperature Range (°C)	State
Crust	5–40	0–870	Solid (rigid) rock
Upper mantle	600	870–2600	Solid (plastic) rock
Mantle	2840	2600–5000	Solid (rigid) rock
Outer core	2200	5000–6400	Liquid
Inner core	1300	6400–6600	Solid

liquid. The inner core has a radius of about 1300 km (800 miles) and is thought to be solid. Both the inner core and outer core are believed to consist of a mixture of iron and nickel. Table 1.1-1 shows the thicknesses, temperatures, and the states of the layers of Earth. *Stop and Think:* How is the structure of a hard-boiled egg similar to the structure of Earth? *Answer:* The eggshell is similar to Earth's crust (both are hard and brittle). The white of the egg compares to Earth's mantle, and the yolk corresponds to Earth's core.

Interesting Facts About How Deep Humans Go into Earth

Gold Mines

The Witwatersrand area in South Africa contains the largest gold resource on Earth. The gold is located deep underground. From Earth's surface, the Western Deep Level Mine shaft plunges about 4 km underground, making it the deepest mine on Earth. No human has descended further into Earth than the miners who work in this mine. In the next few years, the mine may reach a depth of 5 km.

Two problems associated with mining at this depth are high temperature and high rock pressure. Underground temperature in deep mines approach 70°C (158°F). Special cooling systems lower the temperature to 32°C (90°F). The high temperature and the nearly 100 percent relative humidity make working conditions in these mines barely tolerable.

The second problem is caused by the pressure produced by the rock that lies on top of the mine. At this depth, the rock above the mine tunnel produces a pressure of 10,000 tons per square meter. When rock is removed from the mine, the release of great pressure may cause rock bursts. *Rock bursts* are sudden, powerful discharges of large chunks of rock into mine tunnels. Rock bursts kill miners and damage mines.

Activities

1. Cut an apple in half. Observe the cross section of the apple. How is the cross section of an apple like the cross section of Earth?

2. Figure 1.1-2 shows a slice of Earth not drawn to scale. In this activity, you will make a scale drawing of this figure. You will need the following materials: a large sheet of cardboard or paper about 75 × 75 cm (about 30 × 30 inches), a meterstick, a pencil, and a black felt pen. Use Figure 1.1-2 as a reference for making your drawing. The scale of your drawing will be 1:10,000,000—1 cm on your drawing will equal 100 km on Earth. Lightly draw your figure using a pencil before you complete it with the black felt pen.

(a) Draw a 90° angle (a quarter circle). Each side of the angle is 64 cm long. The point of the angle represents the center of Earth.

(b) Draw or trace an arc connecting the open end of the two lines. The arc represents Earth's surface.

(c) Starting at the arc representing Earth's surface, measure down along each side of the angle, placing a mark on each at 4 mm, 6 cm, 29 cm, and 51 cm.

(d) Draw an arc from each mark to the corresponding mark on the other line. The arcs represent boundaries between the layers.

(e) Add all the labels from Figure 1.1-2 on page 5.

SKILL EXERCISE—*Graphing Earth's Internal Temperature*

*E*arth is a sphere of hot material that has been cooling for the past 4.5 billion years. As Earth cooled, a thin crust formed on the surface. Meanwhile, the interior remains hot and slowly releases heat to the surface. The table below shows Earth's temperature at selected depths.

Earth's Temperature at Selected Depths

Location	Depth (km)	Temperature (°C)
Surface	0	10
Mantle	300	1800
Mantle	1000	3200
Mantle	2000	4200
Outer core	3000	5000
Outer core	4000	5800
Outer core	5000	6200
Inner core	6000	6500
Earth center	6400	6600

Questions

1. Copy the graph at the top of the next page into your notebook and plot the data from the table on the graph. Draw a curved line to represent the data.

2. The boundary between the mantle and the outer core exists at a depth of 2900 km. What is the approximate temperature at this boundary?

3. At what depth is the temperature 5500°C?

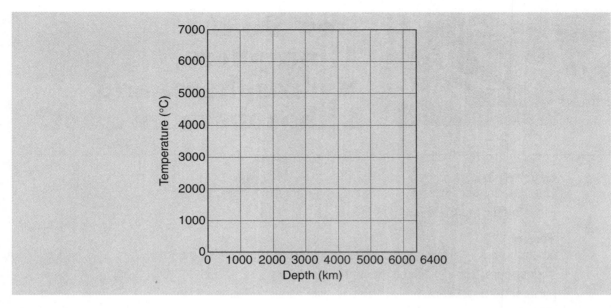

Questions

1. The outermost and thinnest layer of Earth's structure is the
   (1) crust   (3) mantle
   (2) core   (4) hemisphere

2. Comparison: eggshell is to Earth's crust as egg yolk is to
   (1) apple core   (3) apple stem
   (2) Earth's mantle (4) Earth's core

3. The liquid layer of Earth's structure is the
   (1) inner core   (3) outer core
   (2) mantle   (4) upper mantle

4. What layer of Earth's structure has the ability to flow like melting plastic?
   (1) crust   (3) upper mantle
   (2) outer core   (4) inner core

5. Earth's layers formed because the elements that make up Earth differ in
   (1) pressure   (3) temperature
   (2) age   (4) density

Thinking and Analyzing

Base your answers to questions 1-3 on the table at the right and on your knowledge of science. The table shows the density range of Earth's layers.

1. What is the density at the top of the mantle?

2. Describe how density changes as you go deeper into Earth.

3. Explain why the density changes as you go deeper into Earth.

Density Range of Earth's Layers

Location	Density Range (g/cm3)
Crust	2.7–3.0
Mantle	3.3–6.5
Outer core	9.9–12.1
Inner core	12.7–13.0

1.2 How Do the Atmosphere, Hydrosphere, and Lithosphere Interact?

Objectives

Describe the lithosphere, atmosphere, and hydrosphere.

Explain how the three spheres interact.

Terms

lithosphere (LIHTH-oh-sfeer): zone of solid rock that surrounds Earth

atmosphere: zone of gases that surrounds Earth

hydrosphere (HY-droh-sfeer): zone of water that surrounds Earth

inorganic: a substance that is not living and never was living

Earth Is a Sphere and Is Surrounded by Spheres

A *sphere* is a round, ball-like object. A bowling ball, a bubble, and Earth are all *spherical* in shape. Surrounding Earth are three zones called spheres. Since they surround the entire Earth, they have the same shape as Earth, which is why they are called spheres. The solid rock sphere is called the **lithosphere** (*litho-* means "rock"). The gaseous sphere is called the **atmosphere** (*atmo-* means "air"), and the water sphere is called the **hydrosphere** (*hydro-* means "water"). Figure 1.2-1 shows the relationship among the three spheres that surround Earth.

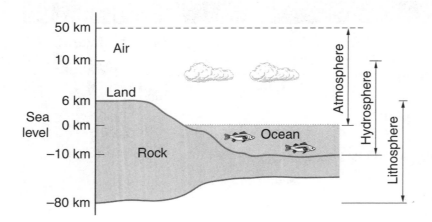

Figure 1.2-1. The atmosphere, hydrosphere, and lithosphere overlap each other near Earth's surface.

Atmosphere

The atmosphere extends upward from Earth's surface to an altitude of about 120 km (74 miles). The air in the upper atmosphere is much less dense than the air in the lower atmosphere. Nitrogen and oxygen are the two major gases that make up the atmosphere. Other gases, including argon, carbon dioxide, and water vapor, are also found in the atmosphere, but in much smaller amounts. All weather takes place in the troposphere, the lowest 18 km (11 miles) of the atmosphere. A layer of *ozone* gas in the upper atmosphere protects all life from the sun's harmful ultraviolet rays.

You might remember from previous science studies that the atmosphere can be divided into several distinct layers based on temperature differences. (See Figure 1.2-2.) ***Stop and Think:*** How does the temperature in the troposphere change as the altitude increases? ***Answer:*** In the troposphere, the temperature decreases as the altitude increases.

Hydrosphere

The hydrosphere consists of all the water on or near Earth's surface. At normal Earth temperatures, water can be found in all three states of matter—solid, liquid, and gas. Water covers about 75 percent of Earth's crust. The oceans contain by far the largest amount of liquid water. Lesser amounts of liquid water can be found in lakes, rivers,

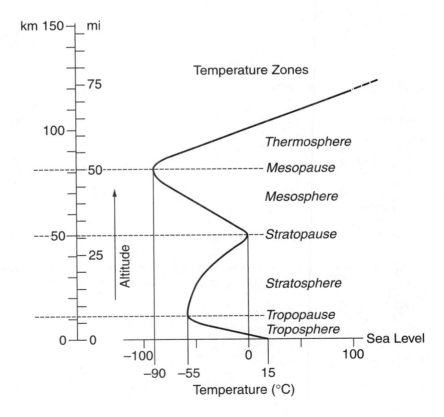

Figure 1.2-2. Layers of the atmosphere are identified by differences in temperature structure.

streams, and underground. Glacial ice in the polar regions and on high mountains contains frozen freshwater. Gaseous water vapor, or humidity, exists throughout the atmosphere. Table 1.2-1 shows how Earth's water is distributed.

Lithosphere

The lithosphere is Earth's solid outer shell. It consists of solid rock in the crust and extends down to the upper mantle, the top part of the mantle. The thickness of the lithosphere varies. The continental lithosphere is about 150 km (about 93 miles) thick, and the oceanic lithosphere is about 70 km (about 44 miles) thick.

The solid rock of the lithosphere, the gases that make up the atmosphere, and the water of the hydrosphere are inorganic. **Inorganic** substances are not living and never were alive.

Table 1.2-1. Distribution of Earth's Water

Source	Percent
Oceans	97.24
Glaciers and icecaps	2.13
Groundwater	0.60
Freshwater lakes	0.01
Inland seas	0.01
Soil/atmosphere/rivers	0.01

Interactions of the Lithosphere, Atmosphere, and Hydrosphere

On Earth's surface, energy is constantly being exchanged among the lithosphere, atmosphere, and hydrosphere. Table 1.2-2 gives examples of energy exchanges among the three spheres.

Table 1.2-2. Examples of Energy Exchange Among Earth's Spheres

Atmosphere → Lithosphere	Physical and chemical weathering by atmospheric gases and moisture cause rocks to crumble.
Atmosphere → Hydrosphere	Winds that blow across water surfaces produce waves. Stronger winds create larger waves.
Lithosphere → Hydrosphere	Earthquakes and volcanic activity on the ocean floor produce tsunamis. These large waves form when strong vibrations in the seafloor transfer energy to the water.
Hydrosphere → Lithosphere	Ocean waves that break along beaches transport sand particles that cause coastline erosion.
Hydrosphere → Atmosphere	Warm ocean currents heat the air above them. The warm air carries heat energy into cooler regions.
Lithosphere → Atmosphere	Volcanic eruptions send ash particles high into the atmosphere. Cooler temperatures occur when these particles block some of the sun's incoming radiation.

SKILL EXERCISE—*Drawing Pie Charts*

*I*n this exercise, you will draw three pie charts representing (1) the composition of the upper lithosphere, (2) the composition of the hydrosphere, and (3) the composition of the lower atmosphere. A pie chart compares the parts of a whole by representing them as sections, or slices of pie. You can easily see what part of the whole pie each piece represents.

Draw three circles in your notebook, tracing around the edge of a small drinking glass to form the outline of each circle. Use a straightedge to draw two light lines that divide each circle into four equal sections. Each section of the circle equals 25 percent. Use the sections in each circle to estimate the size of the pie sections in each pie chart.

1. Use the information in the table below to produce three pie charts. Make one pie chart for each set of data.

2. Give each pie chart a title.

3. Label each section in the three pie charts.

Approximate Chemical Composition (%) of the Upper Lithosphere, Hydrosphere, and Lower Atmosphere

Element	Upper Lithosphere	Hydrosphere	Lower Atmosphere
Oxygen	94.0	33.0	21.0
Nitrogen	–	–	78.0
Hydrogen	–	66.0	–
Other	6.0	1.0	1.0

Activity

Search magazines and travel brochures to find pictures of the lithosphere, atmosphere, and hydrosphere. Cut out each picture and tape it to a page in your notebook. Label each sphere visible in the picture.

Questions

1. Oceans, glaciers, lakes, and rivers are part of Earth's
   (1) atmosphere
   (3) hydrosphere
   (2) hemisphere
   (4) lithosphere

2. Which event represents an energy exchange between the lithosphere and the atmosphere?
   (1) ocean waves
   (2) volcanic eruption
   (3) cloud formation
   (4) radiant energy heating rocks

3. Which of the following is part of the lithosphere?
   (1) elephant
   (3) river
   (2) cloud
   (4) mountain

4. Which sphere contains layers based on temperature differences?
   (1) atmosphere
   (3) hydrosphere
   (2) lithosphere
   (4) crust

5. The lithosphere consists of the
   (1) crust and atmosphere
   (2) crust and hydrosphere
   (3) crust and the top of the mantle
   (4) crust and the bottom of the mantle

Thinking and Analyzing

Base your answers to questions 1 and 2 on the picture below and on your knowledge of science. The picture shows the ocean, rocky sea cliff, and clouds in the sky.

1. What objects in the picture represent the hydrosphere, atmosphere, and lithosphere?

2. Carefully study the picture and describe the interaction between the hydrosphere and atmosphere.

Base your answers to questions 3 and 4 on the passage below and on your knowledge of science.

   Chemical weathering is the breakdown of rocks through changes in their chemical makeup. These changes take place when rocks are exposed to air or water. For instance, when falling rain combines with carbon dioxide in the air, a weak acid called *carbonic acid* forms. Carbonic acid dissolves rocks containing calcite (calcium carbonate), such as limestone and marble.

3. What two spheres interact to form carbonic acid?

4. What two spheres are interacting when carbonic acid dissolves marble rock?

1.3

What Is the Biosphere?

Objectives

Describe the biosphere.

Explain how the biosphere interacts with the other spheres.

Terms

organic: a substance that is living now or was alive in the past

biosphere (BY-oh-sfeer)**:** zone of living things that surrounds Earth

event: change in one sphere that affects another sphere

Biosphere

A fourth sphere also surrounds Earth. Unlike the other three spheres, this sphere is **organic**; it contains living organisms. It is called the biosphere (*bio-* means "life"). (See Figure 1.3-1.) The **biosphere** contains all living things. Like the other spheres described in Lesson 1.2, the biosphere surrounds Earth. In ocean areas, the biosphere extends from the bottom of the ocean upward into the atmosphere. On land surfaces, the biosphere extends from the bottom of the soil to the tops of trees. However, a few organisms exist higher in the atmosphere and lower in the lithosphere. For example, pollen floats high in the atmosphere, and bats live in rock caves underground.

Figure 1.3-1. The biosphere overlaps the atmosphere, hydrosphere, and lithosphere.

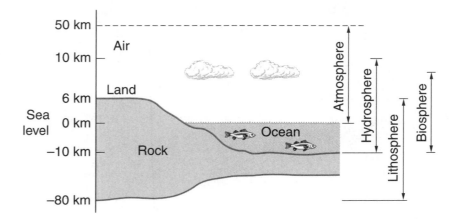

Interaction of the Biosphere and Other Spheres

Figure 1.3-2 shows the relationship among the three inorganic spheres and the biosphere. In Lesson 1.2, we discussed energy and matter exchanges among the three spheres. The biosphere is closely connected to and interacts with the other three spheres. In fact, the biosphere relies on the atmosphere, hydrosphere, and lithosphere for its existence.

A change in one sphere often causes a change in the other spheres. This change is called an **event**. Volcanic eruptions, earthquakes, and hurricanes are sudden and dramatic events. Examples of less dramatic events include rainwater filtering through the soil, or wind blowing grains of sand at the beach. *Stop and Think:* Why is rainwater filtering through the soil called an event? *Answer:* Rainwater filtering through the soil forces the animals that live there out of the soil. This is an example of a change in the hydrosphere causing a change in the biosphere.

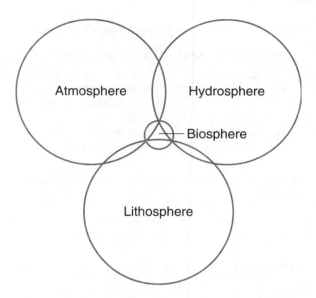

Figure 1.3-2. The relationship among the biosphere, atmosphere, hydrosphere, and lithosphere.

Let us analyze some events to show how the spheres interact with each other. Table 1.3-1 on page 18 explains how an event in one sphere causes changes in other spheres.

Activities

1. Evidence of events between the biosphere and the other three spheres are all around you. Design a chart to record your observations of these events. The chart should have three columns—Column 1: describe the event; Column 2: list the sphere causing the event; and Column 3: list the sphere affected by the event. Identify 10 events and list them in your chart.

2. Use an Internet search engine and look up the term "soil composition." Identify the four major ingredients that are needed to make soil. To what sphere does each of the four ingredients in soil belong? Describe the way plants (biosphere) growing in the soil interact with each of the other ingredients in the soil.

Table 1.3-1. How Events in One Sphere Cause Changes in Other Spheres

Event: Trees destroyed in the rain forest (change in biosphere)

Atmosphere—a decrease in photosynthesis, because there are fewer trees, resulting in less oxygen entering the atmosphere

Lithosphere—removing trees exposes rocks and soil and causes an increase in weathering and erosion

Hydrosphere—a decrease in transpiration, because there are fewer trees, resulting in less water to be put in the hydrosphere

Event: Volcanic eruption releases lava and gases (change in lithosphere)

Biosphere—lava and gases kill local plant and animal life

Atmosphere—ash in the atmosphere blocks sunlight, decreasing the air temperature

Hydrosphere—volcanic eruption puts water vapor in the hydrosphere

Event: Hurricane strikes the land (change in atmosphere)

Biosphere—strong winds and storm surge kill plant and animal life along the coastline

Hydrosphere—heat is moved from water at lower latitudes to water at higher latitudes

Lithosphere—strong winds and storm surge erode the coastline

Event: Formation of acid rain (change in hydrosphere)

Biosphere—acid rain kills the plant and animal life in lakes

Atmosphere—formation of acid rain removes various chemicals from the air

Lithosphere—acid rain increases the weathering of rocks

Interesting Facts About the Interaction of the Biosphere and Atmosphere

The Ozone Layer

There is a layer of ozone gas in the atmosphere at an altitude of 15–55 km (10–30 miles). Ozone, O_3, is a form of oxygen that contains three oxygen atoms instead of two. Ozone is produced by the action of the sun's ultraviolet rays on oxygen atoms in our atmosphere.

The ozone layer acts as a shield protecting life forms from most of the sun's harmful ultraviolet rays. Ultraviolet rays cause sunburn and skin cancer. If all the sun's incoming ultraviolet radiation reached Earth, it would severely damage animal and plant life. The protective layer of ozone is critical to the survival of life on Earth. Human life could not exist without ozone protecting us from ultraviolet radiation. Protecting the biosphere from harmful solar radiation is one important way the biosphere relies on the atmosphere. It is interesting to note that if all the ozone in the upper atmosphere were brought down to Earth's surface, it would form a layer only 0.25 cm (0.1 inch) thick!

Questions

1. Living organisms make up Earth's
   (1) atmosphere   (3) lithosphere
   (2) biosphere   (4) hydrosphere

2. A fish primarily interacts with the
   (1) hydrosphere   (3) atmosphere
   (2) hemisphere   (4) lithosphere

3. An example of trees interacting with the atmosphere is
   (1) absorbing water through their roots
   (2) taking in carbon dioxide and releasing oxygen
   (3) growing leaves
   (4) growing in the cracks of rocks

4. Which event represents an interaction between the biosphere and atmosphere?
   (1) burning wood
   (2) melting ice
   (3) mining iron
   (4) developing of thunderstorms

5. Which statement best describes the location of the biosphere?
   (1) The biosphere is part of the atmosphere.
   (2) The biosphere is part of the hydrosphere.
   (3) The biosphere is part of the lithosphere.
   (4) The biosphere is part of the atmosphere, hydrosphere, and lithosphere.

Thinking and Analyzing

1. Humans are part of the biosphere. Give an example of how humans interact with the
   a) hydrosphere
   b) atmosphere
   c) lithosphere

   Base your answers to questions 2 and 3 on the picture at the right and on your knowledge of science. The picture shows a landscape of an area containing a forest, a stream, and rock exposure.

2. What object(s) in this picture represent the biosphere?

3. What interaction between the biosphere and the atmosphere can you find in this picture?

Review Questions

Term Identification

Each question below shows two terms from Chapter 1. One of the terms is defined.
(1) Choose the term that matches the definition.
(2) Describe how the two terms are different. Following each term is the section (in parenthesis) where the description or definition of the term is found.

1. *Hydrosphere (1.2) — Atmosphere (1.2)*
   The zone of water that surrounds Earth

2. *Inorganic (1.2) — Organic (1.3)*
   Substances that are living or once were living

3. *Mantle (1.1) — Core (1.1)*
   The central portion of Earth

4. *Biosphere (1.3) — Lithosphere (1.2)*
   The zone of life that surrounds Earth

Multiple Choice (Part 1)

Choose the response that best completes the sentence or answers the question.

1. The crust of Earth is part of the
   (1) atmosphere　　(3) hydrosphere
   (2) lithosphere　　(4) mantle

2. Ozone gas in the atmosphere is important because it
   (1) produces rain
   (2) protects us from the sun's ultraviolet light
   (3) helps living things breathe
   (4) produces clouds

3. An energy exchange between the atmosphere and the hydrosphere occurs when
   (1) winds produce waves
   (2) wind blows grains of sand across the ground
   (3) water flows in a stream
   (4) iron rusts in a rock

4. The biosphere interacts with the
   (1) atmosphere
   (2) hydrosphere
   (3) atmosphere and hydrosphere
   (4) atmosphere, hydrosphere, and lithosphere

5. The layer of water that covers most of Earth's surface is part of the
   (1) atmosphere　　(3) lithosphere
   (2) hemisphere　　(4) hydrosphere

6. An event that demonstrates the interaction of the atmosphere and lithosphere is
   (1) the melting of snow
   (2) an oil spill into the ocean
   (3) the rusting of iron in a rock
   (4) the evaporation of ocean water

7. The layers of Earth most probably formed because the elements that make up Earth differ in
   (1) temperature　　(3) pressure
   (2) density　　　　(4) time of formation

8. Which statement is correct?
(1) The outer core is hotter than the inner core.
(2) The outer core is hotter than the mantle.
(3) The mantle is hotter than the outer core.
(4) The crust is hotter than the mantle.

9. Which sphere contains organic substances?
(1) biosphere   (3) atmosphere
(2) hydrosphere   (4) lithosphere

Thinking and Analyzing (Part 2)

1. Base your answers on the picture below.

What object in the picture represents the
(a) atmosphere?   (c) hydrosphere?
(b) lithosphere?   (d) biosphere?

2. Base your responses on the table below and on your knowledge of science. The table lists the percentage of mass for elements in Earth's crust.

Elements	Percent of Mass
Oxygen	46.0
Silicon	28.0
Aluminum	8.0
Iron	6.0
Other	12.0

(a) Draw a circle about the size of a quarter and create a pie chart that represents the data in the table.
(b) Give the pie chart a title.
(c) Label each section of the pie chart with the element and the percentage of mass that it represents.

3. Base your answers to the questions on the figure below and on your knowledge of science. The figure shows three large circles that represent the inorganic spheres, and one small circle that represents the biosphere. The shaded areas represent the interaction of the spheres.

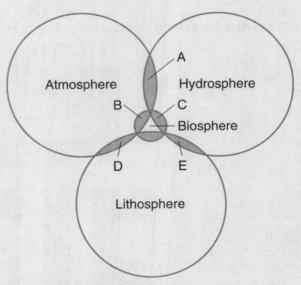

(a) What letter indicates the interaction of the two spheres involved in the formation of a cloud?
(b) What letter indicates the interaction of the two spheres involved in humans breathing?
(c) Describe an event that represents the interaction of the lithosphere and the atmosphere.

4. Base your answers to the questions on the diagram below and on your knowledge of science. The diagram shows the layers of Earth.

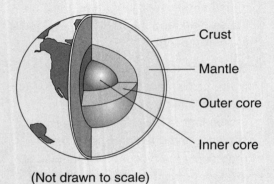

(Not drawn to scale)

(a) Which layer is brittle and breaks, producing earthquakes?
(b) In which layer do you find convection currents within the rock material?
(c) How do the temperature and density change as you go from the crust to Earth's center?

Chapter Puzzle (*Hint:* The words in this puzzle are terms used in the chapter.)

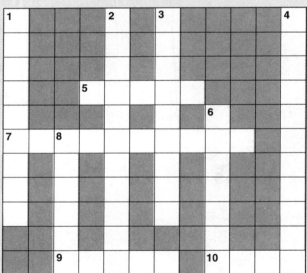

Across

5 outer layer of Earth

7 gaseous zone that surrounds Earth

9 change in sphere that affects other spheres

10 region in Earth's center

Down

1 not living and never was living

2 water zone that surrounds Earth

3 zone of living things that surrounds Earth

4 rock zone that surrounds Earth

6 living or once living substance

8 plastic layer of Earth

Chapter 2

Minerals and Rocks

Contents

Minerals can produce spectacular geometric shapes called crystals.

What Is This Chapter About?

The processes that form rocks and minerals are volcanic activity, sedimentation, and the action of heat and pressure underground. Weathering and erosion are processes that break down rocks and carry the particles away. The big picture of how these processes change rocks is called the rock cycle.

In this chapter you will learn:

1. Minerals are naturally occurring solid substances that are made of inorganic material. Minerals have a definite chemical composition.

2. Rocks are composed of one or more minerals. The three types of rocks are igneous, sedimentary, and metamorphic.

3. Weathering is the breaking down of rocks; erosion is the carrying away of rock particles.

4. The rock cycle illustrates how rock material is reprocessed by a never-ending sequence of events.

Science in Everyday Life

You can find many minerals and rocks around your home and neighborhood. Halite (HAY-lite) is common table salt. Calcite is the white garden stone you see placed around trees or in walkways. Slate is used for floors and walkways. Granite (GRAH-net) and marble are building stones. Jewelry contains mineral gemstones such as diamonds, rubies, and emeralds.

Internet Sites:

http://www.webmineral.com/specimens.shtml See pictures of museum quality minerals.

http://www.bwctc.northants.sch.uk/html/projects /science/ks34/rocks/list.html In this rock identification activity, view pictures of 12 rocks, get the results of various tests, and use charts to identify rocks.

http://volcano.und.nodak.edu/vwdocs/vwlessons /rocks_activities.html This site presents activities and exercises on minerals, magma activity, and volcanic rocks. It focuses on information from the Hawaiian Islands.

2.1

What Are Minerals?

Objectives

Define the term *mineral*.

Describe the physical and chemical properties of minerals.

Explain how to use a flowchart to identify minerals.

Terms

mineral: a naturally occurring solid substance made of inorganic material with a definite chemical composition (chemical formula)

physical property: a characteristic of a mineral that can be observed and measured

chemical properties: how a mineral reacts with other substances

ore: mineral that is mined because it contains metals or other substances of value

crystal: mineral with a regular atomic structure and geometric shape

Minerals

Minerals are naturally occurring solid substances made of *inorganic* (nonliving) material. Each mineral has a definite chemical composition that can be written as a chemical formula. More than 2000 minerals are found in Earth's crust, but only 10 minerals make up 99 percent of the crust. Feldspar is the most abundant mineral. Some other common minerals are quartz, mica, and calcite. Minerals have certain *physical* and *chemical* properties that can be used to identify them.

Physical Properties of Minerals

The characteristics of a mineral that can be observed and measured are called its **physical properties**. Two easily recognized physical properties are color and shape. Other physical properties used to identify a mineral are its streak color, hardness, luster, cleavage, and density.

Color is not always a reliable guide to a mineral's identity. Various samples of the same mineral may have different colors. On the other hand, samples of different minerals may share the same color. Color, as a means of identifying a mineral, is best used after other properties are determined.

Streak color is the color of the powdered form of a mineral. It is observed by scratching the mineral on an unglazed porcelain tile. (See Figure 2.1-1.) **Stop and Think:** Some minerals have no streak color.

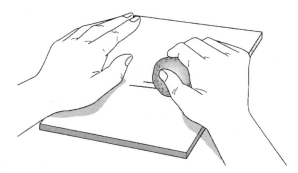

Figure 2.1-1. The streak test is performed by scratching a mineral on the unglazed side of a porcelain tile. A powder form of the mineral is observed on the tile.

Table 2.1-1. Mohs' Scale of Hardness

Mineral	Hardness
Talc	1 (softest)
Gypsum	2
Fingernail	*2.5**
Calcite	3
Penny	*3**
Fluorite	4
Apatite	5
Iron Nail	*5**
Glass	*5.5**
Feldspar	6
Steel File	*6.5**
Quartz	7
Topaz	8
Corundum	9
Diamond	10 (hardest)

Hardness of a common object

Why? **Answer:** A white streak will appear as no streak, and minerals too hard to become powder on a porcelain tile will have no streak.

Hardness is the resistance of a mineral to being scratched. Minerals are assigned a number between 1 and 10 to indicate their hardness, with 1 being the softest and 10 the hardest. Mohs' scale of hardness is shown in Table 2.1-1. It lists minerals for numbers 1-10 as reference points. In addition, it lists the hardness of some common objects used to help identify the hardness of minerals. A mineral can be scratched only by a harder mineral or object, that is, by another mineral or object with a higher number on the hardness scale.

Luster refers to how a mineral looks when it reflects light. Types of luster include metallic, glassy, greasy, and earthy.

Cleavage is a mineral's tendency to break along smooth, flat surfaces. The number and direction of these surfaces are clues to a mineral's identity. Cleavage often causes a

mineral to break into characteristic shapes, as shown in Figure 2.1-2 on page 26. Not all minerals have cleavage; some have *fracture*, which is the tendency to break unevenly.

The *density* of a mineral is calculated by dividing its mass by its volume. The density of liquid water is 1 g/cm³. A mineral's mass can be determined using a triple-beam balance or an electronic balance. Its volume is determined using water displacement. (You learned how to determine the density of an irregular object last year.) Most minerals have a density between 2.5 and 3.5 g/cm³, but some are much higher. Magnetite (iron ore) has a density of 5.2 g/cm³, galena

 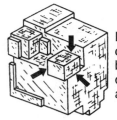

Mica splits into thin sheets because it has one direction of cleavage

Halite breaks into cube-shaped pieces because it has three directions of cleavage at right angles

Figure 2.1-2. Cleavage of mica (left) and halite (right).

(lead ore) has a density of 7.5 g/cm³, and gold has a density of 19.3 g/cm³.

Chemical Properties of Minerals

Minerals also have **chemical properties**, such as how they react chemically with other substances. The mineral calcite has the chemical name calcium carbonate and the chemical formula $CaCO_3$. (See Table 2.1-2.) Calcite, the chief mineral in the rocks limestone and marble, bubbles when a drop of dilute hydrochloric acid (HCl) is placed on it. (See Figure 2.1-3.) The bubbling is caused by a chemical reaction between the calcium carbonate and hydrochloric acid. The chemical reaction releases bubbles of carbon dioxide gas. Most other chemical tests are difficult to administer in the classroom. Table 2.1-2 lists some minerals and their chemical formula.

Table 2.1-2. Chemical Composition of Some Minerals

Mineral	Chemical Name	Chemical Formula
Calcite	Calcium carbonate	$CaCO_3$
Galena	Lead sulfide	PbS
Graphite	Carbon	C
Halite	Sodium chloride	$NaCl$
Hematite	Iron oxide	Fe_2O_3
Quartz	Silicon dioxide	SiO_2

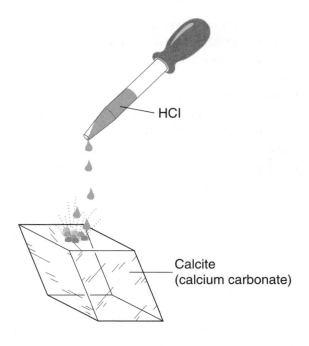

HCl

Calcite (calcium carbonate)

Figure 2.1-3. Calcite produces bubbles when a drop of dilute hydrochloric acid is placed on it.

Table 2.1-3. Some Minerals and Their Uses

Name	Use
Calcite	Cement and garden stone
Diamond and emerald	Jewelry
Feldspar	Ceramics
Galena	Lead products
Garnet	Abrasive (sandpaper) and jewelry
Gold	Jewelry
Graphite	Pencil lead
Halite	Salt
Magnetite and hematite	Iron products
Talc	Talcum powder

Special Properties of Minerals

Sometimes a mineral has a special property that can be used to identify it. For example:

- It's magnetic. Magnetite, a mineral containing a high percentage of iron, is attracted to a magnet.
- The way it feels. Talc is used to make talcum powder. It feels soapy when you rub it between your fingers. Graphite is used to make the lead in your pencil. It feels greasy when you rub it between your fingers. (Do not confuse the lead in a pencil with the metal called lead. They are two different things.)
- How it smells. Flint smells like sulfur (or a burnt match) when it is scratched with a carpenter's nail.

How We Use Minerals

Minerals have a variety of uses. **Ores** are minerals that are mined because they contain metals or some other valuable substance. **Crystals** are minerals with a regular geometric shape. (See Figure 2.1-4.) Some crystals can be cut and polished to make jewelry. Table 2.1-3 lists some minerals and their uses.

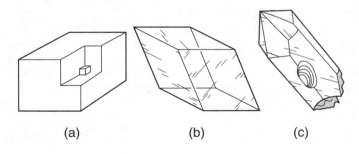

Figure 2.1-4. Crystals of (a) halite, (b) calcite, and (c) quartz.

(a)   (b)   (c)

SKILL EXERCISE—*Identifying Minerals Using a Flowchart*

*E*ach mineral has a set of specific physical and chemical properties that can be used to identify it. Properties are best determined from a mineral that has a fresh, clean surface. If a specimen's surface is dirty or oxidized, it may be necessary to break the specimen to obtain a clean surface. Mineral identification tables organize these properties to make it easier to identify the mineral.

The following steps describe the process of identifying an unknown mineral specimen:

1. Streak color: Scratch the mineral specimen on the unglazed side of a porcelain tile. If a streak appears on the tile, determine the color of the streak. If no streak appears on the tile, the specimen does not have a streak color.

2. Color: Determine the visible color of the specimen. Any shade of black, dark green, brown, red, or blue is considered dark-colored; otherwise the specimen is light-colored.

3. Hardness: Using simple tools such as a glass scratch plate and a carpenter's nail (the hardness of each is between 5 and 6), determine if the mineral is hard (can it scratch the glass plate?) or soft (can it be scratched by the nail?).

4. Cleavage or fracture: This is best determined by breaking the specimen and observing how it breaks. Generally, if the specimen shows flat surfaces, corners, or edges, it has cleavage; otherwise, it has fracture.

5. After you determine the properties of the mineral, use the correct table on page 29 to identify your specimen. If the specimen has no streak, use Table A. If it has a streak, use Table B.
   To use Table A, determine whether the specimen is light- or dark-colored; hard or soft; has cleavage or fracture; and where it fits into the colors and properties listed.
   To use Table B, first determine the streak color. Then decide whether it is hard or soft, has cleavage or fracture, and if it fits into the colors and properties listed.

6. Matching the specimen with known minerals, pictures of minerals, and more extensive descriptions can help assure final identification.

Table A. No Streak

Color	Hard/ Soft	Cleavage/ Fracture	Common Colors/Properties	Name
Light Colored	Hard	Cleavage	Gray, white, pinkish	Feldspar
		Fracture	White, looks waxy	Milky quartz
			Pink, looks glassy to waxy	Rose quartz
	Soft	Cleavage	Clear, salty taste	Halite
			White, soft, scratch with fingernail	Gypsum
			Very soft, soapy feel	Talc
			Colorless, thin sheets peel easily	Muscovite mica
			White to gray, fizzes in weak HCl acid	Calcite
		Fracture	Tan, earthy	Bauxite
Dark Colored	Hard	Cleavage	Black, elongated grains, hardness 5–6	Hornblende
		Fracture	Red, brown, green, looks glassy	Garnet
			Red, brown, yellow, dull or waxy	Jasper
			Black, gray, dull or waxy	Flint
			Gray to black, glassy to waxy	Smoky quartz
	Soft	Cleavage	Black, brown, thin sheets peel easily	Biotite mica
		Fracture	Green to black soapy feel	Serpentine

Table B. Streak Color

Streak Color	Hard/ Soft	Cleavage/ Fracture	Common Colors/Properties	Name
Black/ Dark Green/ or Gray Streak	Hard	Cleavage	No common minerals	
		Fracture	Black, magnetic (magnet sticks to it)	Magnetite
			Brassy yellow, looks metallic	Pyrite
	Soft	Cleavage	Metallic silver, cubes	Galena
		Fracture	Black to gray, marks paper, feels greasy	Graphite
Red/Brown Streak	Hard	Cleavage	No common minerals	
		Fracture	Red to brown, hardness 5–6	Hematite
	Soft	Cleavage	Yellow, brown, looks glassy or waxy	Sphalerite
		Fracture	Yellow-brown, earthy luster	Limonite
			Silver-gray, tiny flakes, looks metallic	Specularite
Green Streak	Soft	Fracture	Bright green, looks earthy, with Azurite	Malachite
Blue Streak	Soft	Fracture	Bright blue, looks earthy, with Malachite	Azurite

Questions

1. You can distinguish between halite and calcite by
   (1) using a magnet
   (2) placing a drop of HCl on each mineral
   (3) observing cleavage
   (4) determining streak color

2. An unknown mineral sample scratches a glass plate but cannot be scratched by a carpenter's nail. The mineral
   (1) is considered hard
   (2) is considered soft
   (3) has a hardness between 5 and 6
   (4) may be graphite

3. The property of the mineral in the diagram that is most easily recognized is _____.

4. Mary has an unknown fresh mineral sample that she wants to identify. She observes the following characteristics of the mineral:
   - It is a dark red-brown color.
   - It produces a reddish streak color.
   - It scratches a glass plate.
   - It does not have corners, flat surfaces, or straight edges.

   What is the name of the mineral?

Activity

A crystal is a mineral with a regular geometric shape. Mineral crystals form from solutions that cool slowly. As the water evaporates out of the solution, the mineral content in the solution slowly forms a geometric shape that is unique and specific to that mineral.

The mineral halite, common table salt, has a cubic shape. Use a magnifying glass to examine some table salt on a dark surface. Can you see that each grain of salt has a cubic shape?

You can produce halite crystals. Dissolve six tablespoons of salt in a shallow dish containing six tablespoons of warm water. Place the solution in a warm place and leave it untouched until all the water evaporates. Examine the salt crystals with a magnifying glass. Can you see the cubic shape of the crystals? Compare these salt crystals with the original crystals.

Questions

1. A chemical property of a mineral is observed if the mineral
   (1) breaks easily when struck with a hammer
   (2) bubbles when acid is placed on it
   (3) is easily scratched by a fingernail
   (4) reflects light from its surface

2. A mineral is a naturally occurring solid substance made of inorganic material. Which of the following is a mineral?
   (1) concrete   (3) brick
   (2) quartz   (4) wood

3. A carpenter's nail has a hardness of about 5. It will scratch
   (1) quartz (hardness 7)
   (2) feldspar (hardness 6)
   (3) diamond (hardness 10)
   (4) calcite (hardness 3)

4. A mineral's tendency to break along smooth, flat surfaces is called
   (1) cleavage   (3) fracture
   (2) luster   (4) hardness

5. Which physical property of a mineral is *least* reliable in identifying a mineral?
   (1) luster   (3) streak color
   (2) cleavage   (4) color

6. The way a mineral reflects light is referred to as
   (1) hardness   (3) color
   (2) luster   (4) density

Thinking and Analyzing

Carefully study the characteristics of the five mineral specimens listed below. Use the mineral identification table on page 29 to identify each one.

Specimen 1: no streak; white color; can be scratched with a fingernail; peels in thin sheets; does not taste salty; does not feel soapy; does not react to acid

Specimen 2: no streak; can be scratched by a carpenter's nail; reacts to acid; white color

Specimen 3: does not react to acid; no streak; white color; scratches glass; breaks producing some flat, smooth surfaces

Specimen 4: no streak; scratches glass; white color; breaks producing an irregular surface

Specimen 5: black streak color; scratches glass; does not have cleavage; a magnet picks it up

How Do Rocks Form?

Objectives
Explain how the three major rock types form.

Describe an example of each type of rock.

Terms
rock: naturally occurring material composed of one or more minerals

igneous (IHG-nee-uhs) **rocks:** rocks that are produced by the cooling and hardening of hot, liquid rock

sedimentary (sehd-uh-MEHN-tuhr-ee) **rocks:** rocks that form from particles called *sediments* that pile up in layers, usually underwater

metamorphic (meht-uh-MAWR-fihk) **rocks:** rocks that are produced when pre-existing rocks undergo a change in form or composition because heat and pressure are applied to them underground

---

Rocks

The **rocks** that form Earth's crust are natural, stony materials composed of one or more minerals. Rocks are identified by their physical and chemical properties. Rocks are classified into three groups—igneous, sedimentary, and metamorphic—depending on how they formed.

Igneous Rocks

Igneous rocks are produced by the cooling and hardening of hot, liquid rock. Melted rock material is produced by underground volcanic activity. Liquid rock underground is called *magma*, and when it reaches Earth's surface it is called *lava*. Volcanic rock that forms underground from magma is called

intrusive igneous rock. When volcanic rock forms from lava that cools at Earth's surface, it is called *extrusive* igneous rock. (See Figure 2.2-1.)

Extrusive igneous rocks cool rapidly on Earth's surface. When lava cools rapidly, tiny grains (also called crystals) usually form. The grains are often too small and too close together to be seen. Basalt is a dark-colored volcanic rock composed of grains too small (less than 1 mm) to be seen with the unaided eye. (See Figure 2.2-2 (a).) Lava on Earth's surface may cool so rapidly that instead of forming individual grains, it forms obsidian, a glassy igneous rock.

Intrusive igneous rocks form from magma underground. (See Figure 2.2-1.) Magma cools at different rates depending on

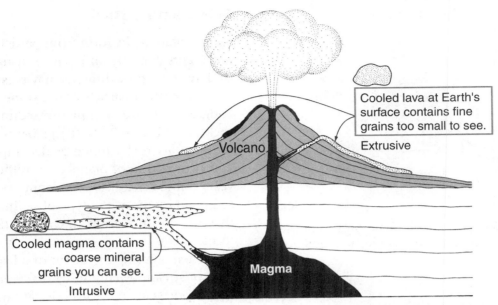

Figure 2.2-1. Volcanic activity produces fine-grained igneous rocks from lava on Earth's surface, and coarse-grained igneous rocks from magma underground.

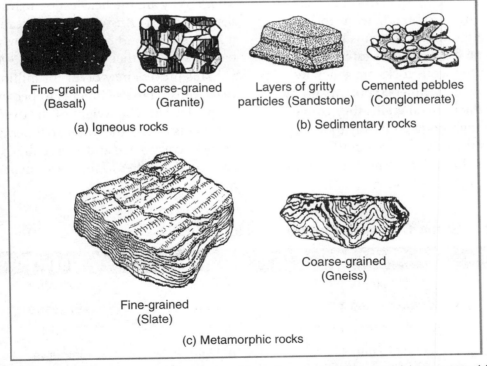

Figure 2.2-2. Examples of (a) igneous rocks, (b) sedimentary rocks, and (c) metamorphic rocks.

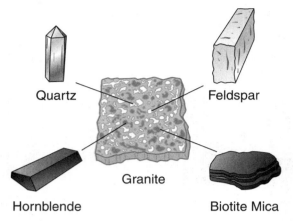

Quartz

Feldspar

Granite

Hornblende

Biotite Mica

Figure 2.2-3. Granite is commonly composed of quartz, feldspar, hornblende, and biotite mica.

how deep it is below Earth's surface. At shallow depths, the magma cools rapidly and small grains form. At deeper depths, magma cools slowly, and larger grains form. Granite is a light-colored igneous rock that forms underground. It contains large (1–10 mm) mineral grains. (See Figure 2.2-3.)

Different igneous rocks are generally identified by their mineral content and by the size of the mineral grains they contain. Color, although not a good way to identify minerals, is the best way to identify very small grains. Table 2.2-1 lists information about some igneous rocks.

Sedimentary Rocks

Sedimentary rocks form from particles called *sediments* that pile up in layers, usually underwater. The sediments may be small rock or seashell particles. Sandstone and conglomerate are examples of sedimentary rocks. (See Figure 2.2-2 (b).) Coal is a sedimentary rock formed by decomposition and compaction of plants over millions of years. Table 2.2-2 lists some common sedimentary rocks, the type of sediments they were formed from, and where they form.

When entering the ocean or a lake, a river or stream usually slows, loses energy, and *deposits* the particles it is carrying. It drops the heaviest particles first, and as the water flow continues to slow, it drops smaller and smaller particles. ***Stop and Think:*** Why are the heaviest particles deposited first? ***Answer:*** As the stream slows, it loses energy and cannot carry the heavier particles any farther. (See Figure 2.2-4.)

This process may occur for millions of years, causing the sediments to become deeply buried. The weight of the overlying sediments causes the lower sediments to become *compacted* and eventually *cemented* together to form sedimentary rock.

Table 2.2-1. Some Common Igneous Rocks

Rock Name	Grain Size	Color	Where Formed
Obsidian	No grains, glassy	Dark	Extrusive
Basalt	Less than 1 mm	Dark	Extrusive
Rhyolite	Less than 1 mm	Light	Extrusive
Granite	1–10 mm	Light	Intrusive
Pegmatite	Greater than 10 mm	Light	Intrusive

Table 2.2-2. Some Common Sedimentary Rocks

Rock Name	Type of Sediment	Place of Formation
Conglomerate	Pebbles or larger	Where fast-running streams slow down
Sandstone	Sand grains	Shallow water, near shore, wave action
Shale	Clay particles	Deep, calm ocean water, lake bottoms
Limestone	Tiny seashell particles	Warm, shallow seas

Metamorphic Rocks

Metamorphic rocks are produced when pre-existing igneous, sedimentary, or metamorphic rocks change their form because of heat, pressure, or both. The process is called *metamorphism* (meht-uh-MAWR-fihz-uhm; *meta-* means "change" and *morph-* means "form"). Metamorphism can occur underground when rocks come in contact with magma, or when forces deep underground squeeze rocks for long periods of time. These high underground temperatures and pressures can change the appearance and mineral composition of the rocks, transforming them into metamorphic rocks.

Marble and slate are metamorphic rocks formed from the sedimentary rocks limestone and shale. Gneiss (pronounced "nice") is a metamorphic rock produced from shale (see Figure 2.2-5 on page 36), a sedimentary rock, or granite, an igneous rock (see Figure 2.2-2c). **Stop and Think:** How do the pressure and temperature change as rocks get buried deeper in the crust? **Answer:** As rocks get buried deeper in the crust, the pressure and temperature increase.

Table 2.2-3 on page 36 lists the pre-existing, original rock, what change takes place in the rock, and the name of the common metamorphic rock it becomes.

Figure 2.2-4.
Sedimentary rocks are produced from sediments deposited in large bodies of water.

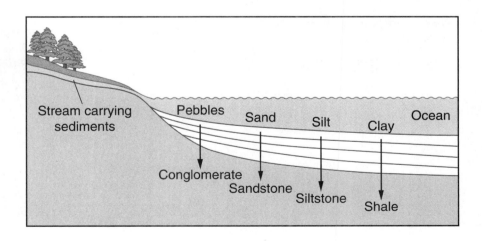

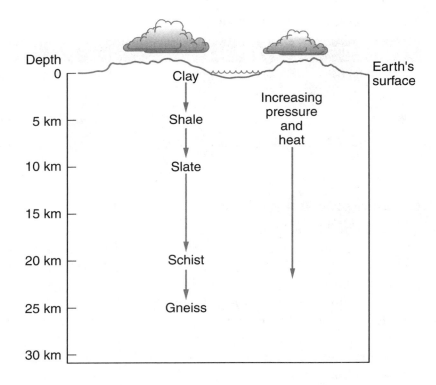

Figure 2.2-5. Shale, a sedimentary rock, can be changed into the metamorphic rock slate, schist, or gneiss, depending upon the amount of heat or pressure or both that is applied.

Table 2.2-3. Some Common Metamorphic Rocks

Original Rock	Change in Rock	Metamorphic Rock Name
Shale (sedimentary)	Compacted, does not easily split	Slate
Limestone (sedimentary)	Harder and denser, crystals form	Marble
Shale (sedimentary) or granite (igneous)	Flattened grains	Gneiss
Sandstone (sedimentary)	Harder and denser	Quartzite

Activities

1. Rocks are commonly used as building materials. Concrete, bricks, and asphalt are not rocks because they are not naturally occurring materials. Carefully observe the rocks in buildings around your home and school and try to identify each.

Design a table in your notebook and record your observations. List the name, use, and a description of each rock you observe. The list below describes the use and description of some common rocks that you might find.

Rock	Use	Description
Slate	Sidewalks and patios	Gray to blue-gray slabs
Granite	Building walls and Gravestones	Orange and black Black and white
Marble	Statues	White with various color streaks
Gneiss	Cobblestones	White-gray with black bands
Sandstone	Garden stones	Various colors, feels like sandpaper

2. Collect a few different rocks about 3–9 cm in size. *Do not take rocks that are on someone else's property.* It may be necessary to break them to find a fresh surface. (CAUTION: ALWAYS WEAR EYE PROTECTION WHEN BREAKING ROCKS.) Use a magnifying glass to help observe the minerals and structure of the rock. Ask yourself the following questions to help identify each rock:

a. Does it contain rounded particles? If it does, it probably is a sedimentary rock. (Sedimentary rocks contain rounded particles.)

b. Does it contain fossils (any part of a plant or animal)? If it does, it is a sedimentary rock. (Fossils are found in sedimentary rocks.)

c. Does it break into layers? If it does, it may be shale (sedimentary) or slate (metamorphic). (Shale and slate break into layers.)

Interesting Facts About Moon Rocks

The first astronauts landed on the moon in 1969. The astronauts brought back 21.7 kg (47.8 pounds) of lunar rocks. Over a period of 3 years, the United States' Apollo astronauts collected and brought back to Earth 342 kg (842 pounds) of moon rock. After much study, the lunar rocks were found to be very much like some Earth rocks, but with a slightly different chemical composition.

Basalt is a dark-colored igneous rock formed from lava on Earth's surface. Samples of basalt were taken from the lunar plains (dark areas on the moon's surface). The basalts on the moon were formed about 3.5 billion years ago when the moon had large lava flows.

Anorthosite (an-ORTH-o-site) is a light-colored igneous rock composed almost entirely of feldspar. Samples of anorthosite were taken from the lunar highlands (bright areas on the moon's surface). The lunar anorthosite was formed about 4.2 billion years ago.

Breccia (BREH-chuh) is a rock composed of angular rock fragments that are cemented together. The breccia on the moon was formed by the impact of large meteors that smashed and battered the moon's surface.

The lunar rocks tell the story of the moon's past. The basalts formed during a period of time when hot lava poured onto the surface of the moon. Breccia formed over billions of years by the impact of meteorites. Because the moon does not have an atmosphere, rain and air have not weathered the rocks. However, the rocks became pitted because of the continuous bombardment of meteorites striking the moon's surface.

Questions

1. Which statement is true of all rocks?
   (1) Rocks form from magma.
   (2) Rocks contain fossils.
   (3) Rocks are composed of minerals.
   (4) Rocks are formed in layers.

2. Igneous rocks are formed by
   (1) weathering
   (2) cementation
   (3) volcanic activity
   (4) sedimentation

3. Metamorphic rocks form from
   (1) sediments
   (2) extrusive volcanic activity
   (3) pre-existing rocks
   (4) intrusive volcanic activity

4. Pegmatite is an igneous rock with mineral grains larger than 10 mm. Granite, another igneous rock, has mineral grains 1-10 mm. A third igneous rock, basalt, has mineral grains less than 1 mm. Which statement is correct?
   (1) Granite formed at the slowest rate.
   (2) Basalt formed at the slowest rate.
   (3) Pegmatite formed at the slowest rate.
   (4) They all formed at the same rate.

5. Sedimentary rocks usually form
   (1) underwater
   (2) from magma
   (3) from pre-existing rocks
   (4) on Earth's surface

Thinking and Analyzing

Base your answers to questions 1-3 on the table below, and the diagram at right, and on your knowledge of science. The table shows the type of sediments that form some common sedimentary rocks. The diagram shows rock particles of different sizes that are carried by a stream and deposited in a lake.

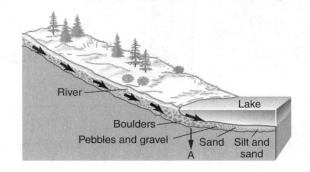

1. Which kind of sedimentary rock in the table has the smallest particles?

2. What rock in the table formed from the sediments at position A in the diagram?

3. What kind of sedimentary rock would form from particles 1.0 mm in size?

Size of Particles	Description	Sedimentary Rock
Less than 0.004 mm	Sheets of tightly packed clay particles	Shale
0.004–0.06 mm	Powdery grains of cemented silt particles	Siltstone
0.06–2.0 mm	Cemented grains of sand	Sandstone
Greater than 2.0 mm	Visible cemented rounded pebbles	Conglomerate

The size of the grains of igneous rocks depends on the rate of cooling or the depth at which the magma cools. The figure below shows the relationships of three igneous rocks, their grain sizes, rate of cooling, and the environment in which the magma became hard.

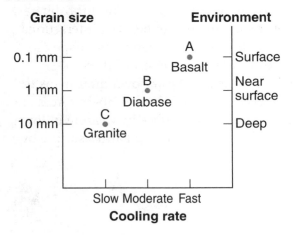

Carefully study the figure and answer questions 4–6 that follow.

4. Which igneous rock has the largest grains?

5. Where do igneous rocks cool the fastest?

6. What is the relationship between the depth at which an igneous rock forms and the size of its grains?

2.3

What Causes the Weathering and Erosion of Rocks?

Objectives

List and describe the agents of weathering.

List and describe the agents of erosion.

Describe the composition of soil.

Terms

bedrock: the solid rock portion of the crust

soil: a mixture of small rock fragments, organic matter, water, and air

weathering: the breaking down of rocks into smaller pieces

erosion (ih-ROH-zhuhn): the process that occurs when particles of rock are removed from the rock and carried away

Bedrock and Soil

Earth's surface consists of bedrock, rock fragments, and soil, as shown in Figure 2.3-1. The agents of weathering and erosion that we will discuss in this lesson slowly change bedrock into rock fragments that may become part of the soil.

Bedrock is the solid rock portion of the crust. Bedrock that is exposed at Earth's surface is called an *outcrop*. Rock fragments are pieces of broken-up bedrock. Fragments can range in size from giant boulders to tiny grains of sand.

Soil is a mixture of small rock fragments and *organic matter* (materials such as decaying leaves and animal wastes). Water

and air are also important parts of soil. Soil and fragments of rock make up most of Earth's land surface, with bedrock hidden underneath.

Changing Earth's Surface

As you learned in Lesson 1.2, the atmosphere, hydrosphere, and lithosphere are constantly interacting. The interactions of the three spheres change the shape of Earth's surface. The effect of these interactions is to wear down mountains to sea level. *Stop and Think:* If these forces are changing Earth's surface by wearing it down, what forces are changing Earth's surface by

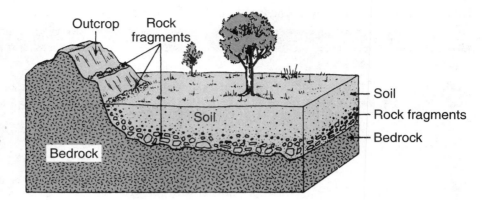

Figure 2.3-1. Earth's surface consists of bedrock, rock fragments, and soil.

lifting it up? *Answer:* Earthquakes and volcanoes lift up Earth's surface.

Weathering

Weathering is the breaking down of rocks into smaller pieces, mainly by water and gases in the atmosphere. Both physical and chemical agents cause the weathering of rocks.

Physical weathering causes rocks to be broken down into smaller fragments. Each rock fragment has the same chemical composition as the larger rock it came from. For example, when water seeps into a crack in a rock, the water may freeze. As the water freezes, it expands and breaks the rock, as shown in Figure 2.3-2. Plant roots that grow in the cracks in rocks can also split rocks apart.

Chemical weathering is the breaking down of rocks through changes in their chemical makeup. These changes take place when rocks are exposed to air or water. For instance, when falling rain combines with carbon dioxide in the air, a weak acid called *carbonic acid* is formed. Carbonic acid dissolves rocks containing calcite (calcium

carbonate), such as limestone and marble. Also, when oxygen and water react chemically with rocks that contain the mineral iron, the iron changes into rust, which crumbles away easily. Table 2.3-1 on page 42 lists agents (causes) of physical and chemical weathering, and provides an example of each.

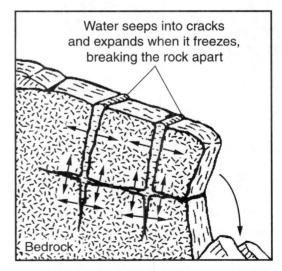

Figure 2.3-2. Physical weathering caused by water freezing in rock cracks.

Table 2.3-1. Agents and Examples of Weathering

Agent of Weathering	Description and Example
Physical Weathering	
Ice wedging	When water freezes in a rock crack, the water expands and splits the rock. Roadside rocks in a falling rock zone after a winter thaw are examples of the results of ice wedging.
Water abrasion	Sand carried by running water scrapes rocks in a stream. This process produces smooth and rounded rocks.
Wind abrasion	Blowing sand scrapes rock surfaces. Smooth surfaces on rocks in dry regions are formed by wind abrasion.
Exfoliation	When pressure is removed from rock that formed deep underground, thick slabs tend to peel off the surface. This is commonly seen in granite at the top of mountains.
Plant roots	Trees and plants growing out of cracks in large rocks split the rock apart.
Animals	Burrowing animals bring rocks to the surface. The rocks are then exposed to the other agents of weathering.
Chemical Weathering	
Carbonic acid	Carbonic acid in rainwater dissolves limestone and marble. Underground caverns are produced when rock is dissolved by acidic water.
Oxidation	Oxygen reacts with some materials, changing them chemically. When oxygen reacts with rocks containing iron, a red stain forms on the rock.
Plant acids	Acids are produced when organic matter decays in water. Some minerals react with acids, causing the mineral to crumble.
Water	Some minerals react with water. Water causes feldspar to decompose and produce clay.

Erosion

Erosion is the process by which rock particles are removed from rocks and carried away. Erosion requires a moving force, such as running water, which can carry the rock particles. Gravity provides the force that moves water and rock material downhill.

The process of erosion can be observed after a heavy rain, when streams turn muddy brown from the rock material in the water. Erosion and gravity are constantly at work, slowly moving rock material from higher elevations to lower elevations. The effects of gravity and erosion produce very visible land features along hillsides where landslides,

Figure 2.3-3. Erosion caused by running water carved the Grand Canyon, a canyon more than a mile deep.

mudslides, and avalanches have occurred. These features are commonly seen after heavy rain or snow in mountainous regions.

Running water in rivers and streams also carves out valleys and canyons. The Grand Canyon in Arizona is a spectacular example of erosion caused by running water. (See Figure 2.3-3.)

Table 2.3-2 lists the agents of erosion and an example of each.

Table 2.3-2. Agents and Examples of Erosion

Agent of Erosion	Example of Erosion
Gravity	Landslides and mudslides carry rock and soil downhill.
Running water	Rock particles are carried in the running water.
Groundwater	Rock material is removed underground to form caverns.
Glaciers	Rock material is carried under and in a glacier.
Winds	Rock particles are carried and rolled along near the ground.
Longshore currents	Sand is moved along a beach by waves washing in and out.

SKILL EXERCISE—*Predicting an Experimental Result*

Rocks in a stream constantly knock and scrape against each other and against larger rocks as they are carried along by flowing water. This is called stream abrasion. The longer the rocks are in the stream, the more they tumble about and strike one another. To demonstrate this action and study its effects, a student carried out the following experiment.

Twenty-five marble chips and a liter of water were placed in a large coffee can marked A. The can was covered with a lid and shaken for 30 minutes (by many students). Then another 25 marble chips and a liter of water were placed in a second can, marked B, and covered. This can was shaken for 120 minutes (by even more students). The illustration below shows the materials used in the experiment. Keep in mind what you have learned about weathering to help you answer the following questions.

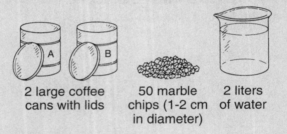

2 large coffee   50 marble   2 liters
cans with lids   chips (1-2 cm of water
                 in diameter)

Questions

1. Which of the following best predicts the result of the experiment?

(a) The marble chips in can A will be smaller and rounder than the chips in can B.

(b) The marble chips in can B will be smaller and rounder than the chips in can A.

(c) There will be no difference between the marble chips in cans A and B.

2. Which graph below best predicts what would happen to rocks in a fast-moving stream over time?

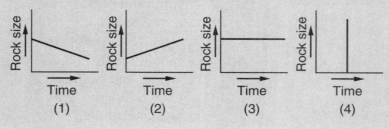

(1) (2) (3) (4)

Activity

Find five or more pictures of weathering and erosion features. Travel magazines and brochures are usually good places to look. Also, look at advertisements in various magazines. Tape the pictures in your notebook. Write a sentence describing the type of weathering or erosion each picture shows.

Interesting Facts About Glaciers

About 100,000 years ago, a continental ice sheet advanced from Canada and reached across the United States from New York City to Seattle, Washington. In the Midwest it extended as far south as Kansas and Missouri. In some places, the glacier was a mile or more thick, but by the time it reached its most southern point, it was thinner because of the warmer climate.

The glacier picked up, pushed, and carried lots of sand and gravel. Eventually, it deposited the rock material along its leading edge, forming a ridge called a *moraine*. The terminal moraine marks the most southerly extent of the glacier, and provides evidence of just how far south the glacier moved,

About 15,000 years ago, the glacier started to melt and retreat. The coastline along the east and west coast changed dramatically, as sea level rose about 125 meters (400 feet) during the next 5,000 years. It is hard to imagine that about 15,000 years ago ice covered Chicago perhaps to a height equal to that of the Sears Tower!

Questions

1. When water freezes in rock cracks, the ice expands and the rock may break apart. This is a type of
   (1) glacial erosion
   (2) physical weathering
   (3) chemical weathering
   (4) groundwater erosion

2. The diagram below shows the mineral magnetite, an ore of iron, changing into rust particles. This is an example of

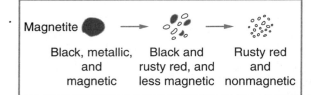

Magnetite

Black, metallic, and magnetic　Black and rusty red, and less magnetic　Rusty red and nonmagnetic

   (1) physical weathering
   (2) chemical weathering
   (3) erosion by running water
   (4) the role of gravity in erosion

3. Erosion is the process by which rocks at Earth's surface
   (1) are removed and carried away
   (2) crumble and decay
   (3) turn into rust
   (4) melt to form magma

4. Two important agents of erosion are
   (1) volcanoes and earthquakes
   (2) water and carbonic acid
   (3) gravity and running water
   (4) ice wedging and plant roots

5. A mixture of rock fragments, organic matter, water, and air is called
   (1) bedrock
   (2) soil
   (3) an outcrop
   (4) the crust

Thinking and Analyzing

Base your answers to questions 1 and 2 on the diagrams below and on your knowledge of science. The diagrams show what a land area looked like 50 million years ago (A) and 10 million years ago (B).

1. What geologic processes have changed the land?

2. If the processes continued to act on the land, predict what the land would look like today.

3. Identify each example below as a feature of weathering or erosion, and name the agent that caused it.
   (a) a rounded rock in a stream bed
   (b) a stream carrying muddy water
   (c) a tree growing from a crack in a large rock outcrop
   (d) a carved marble gravestone that is difficult to read

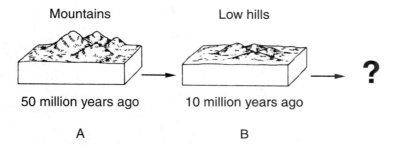

Mountains   Low hills

50 million years ago   10 million years ago

A   B

2.4 What Is the Rock Cycle?

Objectives

Describe processes that change rocks.

Describe the rock cycle.

Term

rock cycle: the never-ending series of natural processes that change one type of rock into another type of rock

---

The Rock Cycle

In Lesson 2.2, we discussed the three processes that form rocks. In Lesson 2.3, we discussed the processes of weathering and erosion that break down rocks and carry rock particles away. In this lesson, we will discuss how rock-forming and rock-breaking processes are related, and how they can repeat many times. The never-ending series of natural processes that change one type of rock into another type of rock is called the **rock cycle.** (See Figure 2.4-1.)

The three types of rocks—igneous, sedimentary, and metamorphic—can be transformed into new types of rocks by natural processes. The weathering and erosion of pre-existing rocks can produce sediments that can be deposited, buried, and cemented into new sedimentary rocks. Heat or pressure or both deep within Earth's crust can also affect pre-existing rocks, transforming them into new metamorphic rocks. The heat of volcanic activity can melt pre-existing rocks into a *magma* that can *solidify* (harden) into new igneous rocks.

Examples of the Rock Cycle

The rock cycle is a slow process because the processes involved in weathering and erosion, depositing and burying sediments, and melting rock are slow. Therefore, the processes involved in the rock cycle may take millions, or even billions, of years.

Let us use Figure 2.4-1 to help us list the steps involved in the changing of an igneous rock into a sedimentary rock.

Igneous Rock → Sedimentary Rock
1. Igneous rock
2. Weathering and erosion
3. Sediments
4. Deposition
5. Burial and compaction
6. Cementation
7. Sedimentary rock

Stop and Think: How might an igneous rock change into a metamorphic rock?

Figure 2.4-1. The rock cycle illustrates the never-ending series of natural processes that change one type of rock into another.

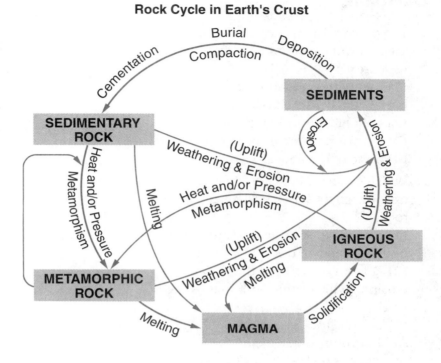

Rock Cycle in Earth's Crust

Answer: By referring to Figure 2.4-1, we can come up with the following steps:

1. Igneous rock
2. Heat or pressure or both
3. Metamorphism
4. Metamorphic rock

Activity

Make a large display of the rock cycle.

1. Igneous rock: Draw or glue a small igneous rock to the display.
2. Sedimentary rock: Draw or glue a small sedimentary rock to the display.
3. Metamorphic rock: Draw or glue a small metamorphic rock to the display.
4. Sediments: Draw or glue sediment particles to the display.
5. Magma: Draw or tape a picture of an erupting volcano.
6. Weathering and erosion: Draw or tape a picture of a weathering process and a picture of an erosion process to the display.
7. Label all the other processes.

Questions

1. The processes necessary to change a metamorphic rock into a sedimentary rock are
   (1) weathering and erosion, deposition, burial, and cementation
   (2) melting and solidification
   (3) heat or pressure or both
   (4) none of the above

2. Which of these sequences changes a pre-existing igneous rock into a new igneous rock?
   (1) igneous rock → heat or pressure or both → igneous rock
   (2) igneous rock → weathering and erosion → deposition → burial → igneous rock
   (3) igneous rock → melting and solidification → igneous rock
   (4) igneous rock → heat or pressure or both → metamorphism → igneous rock

3. Using Figure 2.4-1 on page 49, determine how many years it takes for an igneous rock to become a sedimentary rock.
   (1) 100 years
   (2) 500 years
   (3) 1,000,000 years
   (4) cannot be determined

4. What processes are involved in changing a sedimentary rock into a metamorphic rock?
   (1) heating and melting
   (2) weathering and erosion
   (3) burial and cementing
   (4) heating or pressure or both

5. Magma will solidify into
   (1) an igneous rock
   (2) a metamorphic rock
   (3) a sedimentary rock
   (4) sediments

Thinking and Analyzing

Base your answers to questions 1–3 on Figure 2.4-1 on page 49. The figure represents the rock cycle.

1. What process changes metamorphic rock into magma?

2. Give an example of a weathering agent and an erosion agent that might cause metamorphic rock to become sediment.

3. List the steps that can cause a sedimentary rock to change into a metamorphic rock.

4. The table below lists the three types of rocks and their method of formation. Copy the table and complete it by filling in the missing items (A–C).

Rock Type	Method of Formation
(A)	melting and solidification
(B)	deposition, compaction, and cementation
Metamorphic	(C)

Review Questions

Term Identification

Each question below shows two terms from Chapter 2. One of the terms is defined. (1) Choose the term that matches the definition. (2) Describe how the two terms are different. Following each term is the section (in parenthesis) where the description or definition of the term is found.

1. *Mineral (2.1) — Rock (2.2)*
   A naturally occurring, solid substance made of inorganic material with a definite chemical composition

2. *Quartz (2.1) — Calcite (2.1)*
   A soft mineral that produces bubbles when hydrochloric acid is placed on it

3. *Granite (2.2) — Basalt (2.2)*
   An intrusive igneous rock that forms from magma

4. *Magma (2.2) — Lava (2.2)*
   Hot liquid rock on Earth's surface

5. *Igneous rock (2.2) — Sedimentary rock (2.2)*
   Rock that usually forms by particles settling underwater

6. *Weathering (2.3) — Erosion (2.3)*
   The process that removes and carries away rock particles

7. *Rock cycle (2.4) — Metamorphism (2.2)*
   The never-ending series of natural processes that change rocks

Multiple Choice (Part 1)

Choose the response that best completes the sentence or answers the question.

1. Exposed bedrock on Earth's surface is called a(n)
   (1) mountain   (3) outcrop
   (2) boulder   (4) rock fragment

2. A chemical property that helps to identify a mineral is
   (1) luster   (3) reactivity to acid
   (2) hardness   (4) cleavage

3. The hardness of six minerals is listed in the table below.

Hardness of Some Minerals

Mineral	Hardness
Talc	1
Mica	2
Calcite	3
Fluorite	4
Hornblende	6
Garnet	7

Which mineral is hard enough to scratch fluorite but will not scratch garnet?

(1) mica   (3) hornblende
(2) calcite   (4) talc

4. Rocks that form from layers of small particles are called
   (1) metamorphic rocks
   (2) sedimentary rocks
   (3) igneous rocks
   (4) volcanic rocks

5. *Schist* is a metamorphic rock found in New York's Central Park. Schist formed by
   (1) cooling and hardening of magma underground
   (2) great heat or pressure or both underground
   (3) accumulation of sand grains in a lake
   (4) accumulation of pebbles in a lake

6. *Granite* and *rhyolite* are igneous rocks. They have the same chemical composition and form from the same liquid rock solution. Why are the crystals in granite larger than the crystals in rhyolite?
   (1) Granite cools more slowly than rhyolite.
   (2) Rhyolite cools more slowly than granite.
   (3) Granite forms closer to shore than rhyolite.
   (4) Rhyolite forms from sediments.

7. The grains in igneous rocks are
   (1) cemented sediments
   (2) formed from pre-existing rocks
   (3) organic
   (4) individual minerals

8. The mineral calcite and the metamorphic rock marble produce bubbles when a drop of hydrochloric acid is placed on them. This tells us that calcite and marble have the same
   (1) hardness
   (2) color
   (3) chemical composition
   (4) physical properties

9. Coal is a naturally occurring substance that formed from the decomposition and compaction of plants over millions of years. Coal is a
   (1) mineral formed underground
   (2) mineral formed on the surface
   (3) rock
   (4) man-made substance

10. Under the soil is the solid rock portion of the crust. This rock portion of the crust is called the
    (1) bedrock
    (2) lower rock
    (3) lithosphere
    (4) mantle

11. After a heavy rain, Bea noticed the stream near her house was brown and muddy. Muddy water is an example of
    (1) erosion
    (2) physical weathering
    (3) chemical weathering
    (4) rusting

12. What is formed when falling rain mixes with carbon dioxide in the air?
    (1) rust
    (2) clay
    (3) carbonic acid
    (4) hydrochloric acid

Thinking and Analyzing (Part 2)

Base your answers to questions 1-3 on the diagram below and on your knowledge of science. The diagram shows a portion of the rock cycle.

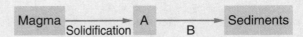

1. What kind of rock is represented at position A?

2. What two processes occur at position *B*?

3. What are two other ways that rock type *A* could change?

4. Base your answers to the questions on the diagram and table below. The most common elements in Earth's crust are shown in the pie chart, and the table lists the name and symbol of each element.

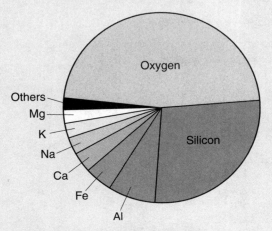

Element	Symbol
Oxygen	O
Silicon	Si
Aluminum	Al
Iron	Fe
Calcium	Ca
Sodium	Na
Potassium	K
Magnesium	Mg
All others	

(a) About what percent of Earth's crust is oxygen and silicon?

(b) Explain how the pie chart tells us that there is more aluminum than iron in Earth's crust.

(c) Feldspar is the most abundant mineral in Earth's crust. There are different types of feldspar. One type has the chemical composition $KAlSi_3O_8$. What elements are in this feldspar?

5. For each of the following descriptions, state whether it is an example of weathering or erosion, and what agent caused it.

(a) Ice hanging out of a rock exposed along a roadside

(b) Grass and bushes growing between the cracks in a rock

(c) Sand blowing in your face

(d) Caverns formed in limestone

6. The table below lists the three rock types. Copy the table into your notebook and fill in the right side with the type of material from which each rock formed.

Rock	Formed From
Igneous	(a)
Sedimentary	(b)
Metamorphic	(c)

Chapter Puzzle (*Hint:* The words in this puzzle are terms used in the chapter.)

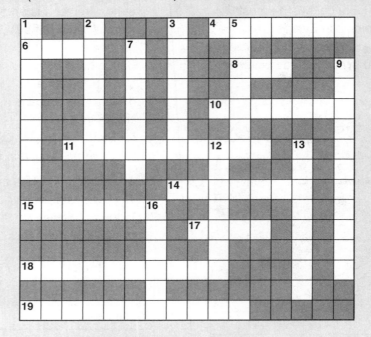

Across

4 solid rock portion of crust

6 liquid rock on Earth's surface

8 mined valuable mineral

10 gem with hardness of 10

11 breaking down of rocks into smaller pieces

14 examples: feldspar, quartz, and calcite

15 mineral with regular geometric shape

17 mixture of rock fragments, organic matter, water, and air

18 minerals have physical and chemical _____

19 type of rock formed by heat or pressure or both

Down

1 tendency to break along smooth flat surfaces

2 bubble reaction to hydrochloric acid

3 light-colored intrusive igneous rock

5 rock particles removed and carried away

7 force that moves water and particles downhill

9 type of rock formed by particles settling in water

12 type of rock formed from magma

13 most abundant mineral in crust

16 how a mineral reflects light

Chapter 3

Fossils and Earth's History

Contents

It is now believed that the extinction of the dinosaurs was caused by the impact of an asteroid.

What Is This Chapter About?

For about 4.5 billion years, the surface of planet Earth has slowly changed. Geologic events such as earthquakes, volcanoes, and erosion have been recorded in the formation and destruction of rocks. Biological events such as animal evolution and extinction have been recorded in fossils. Geologists (scientists who study Earth) study rocks and fossils to understand the history of planet Earth.

In this chapter you will learn:

1. Fossils are the remains or traces of ancient life forms.

2. Rocks can be dated relative to geologi and biological events.

3. A rock's age in years can be determined by using radioactive dating methods.

4. The geologic time line lists geologic and biological events in the order they occurred in time.

Science in Everyday Life

Do you want to see a fossil? Fossils are found in sedimentary rocks. If you are lucky enough to live in an area that has sedimentary rocks and the rocks contain fossils, you can observe the fossils in their natural state. You may even want to start a fossil collection. However, many areas in the United States do not contain fossils in nearby rocks. For you to see fossils, you may have to go to a local museum.

Internet Sites:

http://www.enchantedlearning.com/subjects/dinosaurs/ Do you want to learn more about dinosaurs? This site contains lots of information about dinosaurs and fossils.

http://www.fossilmuseum.net/index.htm The Virtual Fossil Museum has information about geologic time and fossils. It also has many pictures of fossils.

http://www.amnh.org/exhibitions/dinosaurs/extinction/ Learn about what happened to the dinosaurs 65 million years ago. Explaining the mass extinction of the dinosaurs and other organisms has presented a problem to scientists for a long time. They believe they finally have the answer to what caused the mass extinction.

3.1

What Are Fossils?

Objectives

Describe how fossils form.

Explain what fossils tell us about Earth.

Terms

fossils: the remains or traces of organisms that lived long ago

mold: a hollow area in sedimentary rock produced where a fossil dissolved out of the rock; an imprint of a fossil

cast: a mold filled with new mineral matter

index fossil: the fossil of an animal or plant that existed for a relatively short time and lived over a wide area, used to match and date rock layers

Fossils

Fossils are the remains or traces of organisms that lived long ago. Fossils usually form when a dead plant or animal, or some trace of it, like a footprint in mud, is covered by sediment that slowly compacts and hardens into rock. Fossils are found in sedimentary rocks. The heat and pressure associated with igneous and metamorphic rocks usually destroy all organic remains and make it difficult to produce a fossil in these rocks. Figure 3.1-1 shows examples of some common fossils.

How Fossils Form

Rare conditions must exist in order for fossils to be preserved. Fossils commonly form when the remains of an organism are buried in sediment soon after the organism dies. Soft, fleshy parts of an organism usually decay and rarely become fossils. Harder parts of animals, such as bones, shells, and teeth, are more likely to become fossils. Some of the ways organisms are preserved as fossils include the following:

- *Original remains.* In rare cases, the actual remains of the original animal are preserved. The woolly mammoth, an elephant-like animal, lived in Siberia 10,000–20,000 years ago during the last ice age. Complete remains of this animal, including flesh, hair, and fur, have been found preserved in ice and frozen ground. (See the feature, Interesting Facts About Woolly Mammoths, on page 58.)

- *Replacement.* In many cases, the original organism's remains were

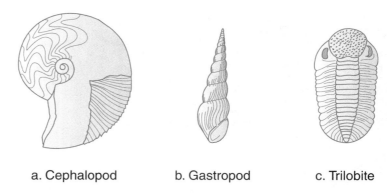

a. Cephalopod   b. Gastropod   c. Trilobite

Figure 3.1-1. Examples of some common fossils.

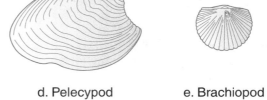

d. Pelecypod   e. Brachiopod

buried, decayed, and slowly removed by groundwater. As the organism decayed, minerals in the groundwater replaced the organic material. For example, petrified trees formed when trees were buried in sediment. As the trees decayed, the wood was slowly replaced with a mineral in groundwater called silica, a form of quartz.

- *Molds and casts.* Often, a preserved shell in sedimentary rock dissolves, leaving a hollow area called a **mold**. Sometimes the mold is filled with new minerals and a **cast** is produced. Shellfish fossils such as brachiopods (BRAKE-ee-oh-podz) and pelecypods (pe-LEH-seh-podz) are commonly found as molds and casts. (See Figure 3.1-1.)
- *Impressions.* Footprints and impressions of whole organisms that

were produced in wet sand or clay may be preserved when the sediments become rocks. Fossil impressions of dinosaur footprints, burrowing worms, and plant parts have been found in sandstone and shale.

What the Fossil Record Tells Us

Geologists have learned much about Earth's past by studying fossils. Fossil evidence has helped geologists trace the evolution of life on Earth. When fossils are arranged in order from oldest to youngest, they indicate a progressive evolution of life through time. The fossil record also reveals periods of time when some species became extinct, and when some species suddenly flourished and became widespread.

Fossils provide clues to ancient environments. For example, coral lives only

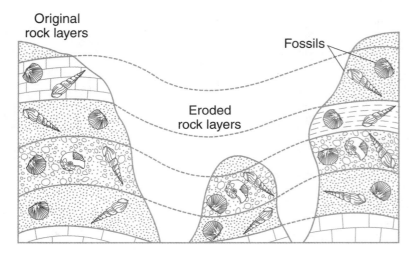

Figure 3.1-2. Fossils can be used to match distant rock locations, regardless of rock type.

in warm waters with lots of sunlight. An area with a lot of fossil coral suggests that it was once covered by a warm, shallow sea.

Fossils can sometimes be used to match rock layers that are located far away from each other. Geologists match rock layers to better understand past environments. (See Figure 3.1-2.) Rock layers from different locations commonly have different physical features, even though the rocks may have formed at the same time. Fossils are a reliable method of determining if rock layers formed at the same time. This is especially true if the animal or plant species existed for a relatively short time and lived over a wide area. This type of fossil is known as an **index fossil**. It provides an excellent means of matching the relative age of two distant rock formations.

Interesting Facts About Woolly Mammoths

Woolly mammoths, elephant-like creatures, became extinct thousands of years ago. During the last ice age, they roamed in herds near the leading edge of the ice sheet. They became extinct because of climate changes, and probably because humans hunted them for food.

In 1900, a Russian hunter in Siberia discovered a woolly mammoth in the frozen ground. The woolly mammoth, unlike the elephant that lives much farther south today, had long curved tusks and a body covered with coarse hair and woolly fur. The discovered animal contained well-preserved flesh and was nearly complete.

Since this discovery in Siberia, other frozen woolly mammoths have been found in Siberia and Alaska. Fossils containing soft fleshy parts are rare and occurred because the animal was frozen quickly. This process of preserving woolly mammoths is the same as preserving steaks in your home freezer!

Activities

1. Thousands of fossil bones from the dinosaur Coelophysis (SEE-low-feh-zis) have been discovered in New Mexico, Arizona, and Utah. A small number of fossil footprints have also been found in New York State. The fossils are found in rocks that were deposited about 200 millions years ago. Do some research on the Internet or in the library and answer the following questions about *Coelophysis*:

(a) In what type of rock were the footprints found?

(b) How large was Coelophysis?

(c) Coelophysis was

    (1) an herbivore   (3) an omnivore

    (2) a carnivore   (4) a producer

2. Make a fossil. Cover a clamshell (3–6 cm long) with petroleum jelly. Push it into a flattened piece of clay that is at least 2 cm thick. Carefully pull the shell out of the clay. Cut a strip of cardboard 3 cm wide and long enough to fit around the shell impression. Press the cardboard into the clay to form a loop around the impression. Cover the clay inside of the cardboard with petroleum jelly. Mix plaster of paris and pour it inside the loop of cardboard over the shell impression. After the plaster dries, remove it and paint it to look like a rock. What type of fossil preservation does the clay and plaster fossil represent?

Questions

1. Geologists can use fossils to
   (1) match rock layers at different places
   (2) study the evolution of life
   (3) determine past environments
   (4) all of the above

2. Fossils are found in
   (1) igneous rocks
   (2) sedimentary rocks
   (3) metamorphic rocks
   (4) volcanic rocks

3. Mary found a coral fossil in a limestone outcrop near her home. Which statement about the area's past is most likely correct?
   (1) The area was once under deep, cold water.
   (2) The area was never underwater.
   (3) The area was once under cold, shallow water.
   (4) The area was once under warm, shallow water.

4. An animal fossil has the best opportunity to form if the animal is
   (1) buried slowly and contains soft parts
   (2) buried slowly and contains hard parts
   (3) buried quickly and contains soft parts
   (4) buried quickly and contains hard parts

5. We know that rock layers at two locations 100 km apart most likely formed at the same time because they have
   (1) rocks that are similar colors
   (2) the same thickness
   (3) similar fossils
   (4) the same depth below the surface

Thinking and Analyzing

1. Base your answers to the questions on the fossil table at the right and on your knowledge of science. The fossil table shows when three fossils lived. Assume that if a fossil is not shown, it did not exist at that time.
   (a) About how many years did trilobites (TRY-low-bytes) exist?
   (b) What happened to the brachiopods 144 million years ago?
   (c) A rock was found that contained trilobites and brachiopods but not cephalopods (SEF-ah-lo-podz). When did the rock form?

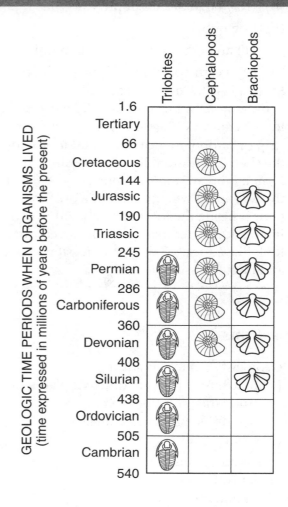

2. Base your answers to the following questions on the diagram at the right and on your knowledge of science. Columns A, B, and C represent layers of rock at three different locations. Within the rock layers are fossils: gastropods (GAS-troh-podz), trilobites, brachiopods, and horn coral.

(a) What type of rock is found at all three locations?

(b) Which fossil is the best index fossil?

(c) The trilobite fossil is a cast. Explain how a cast forms.

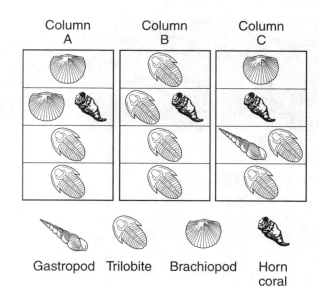

Column A   Column B   Column C

Gastropod   Trilobite   Brachiopod   Horn coral

3.2 How Do You Determine the Relative Age of Rocks?

Objectives

Describe the laws and principles used to determine the relative age of rocks.
Determine the relative age of rocks and events in a rock diagram.

Term

relative age: the order or sequence in which rocks formed

Interpreting Rocks

The **relative age** of rocks refers to the order, or sequence, in which rocks form. Geologists are able to determine the relative age of rocks and interpret geologic events. They do this using the following laws and principles of geology:

- Younger sedimentary rocks are deposited over older sedimentary rocks. When sedimentary rocks form underwater, the oldest rocks are on the bottom and the youngest rocks are on top. If mountain-building forces, such as earthquakes and volcanoes, do not overturn the rocks, they remain in this position. (See Figure 3.2-1.)
- Geologic processes occurring today are the same geologic processes that have occurred throughout Earth's history. Scientists know that the forces that produce earthquakes and volcanoes today are the same forces that have bent, tilted, and produced massive cracks in rocks during the entire

geologic history of Earth.
- A buried erosion surface represents a period of time when the land was above sea level and erosion removed some rock layers. Mountains are produced when Earth forces cause land to uplift. When mountains erode and are covered with younger layers of rock, a buried erosion surface is produced. An erosion surface represents an interruption in rock-forming events.

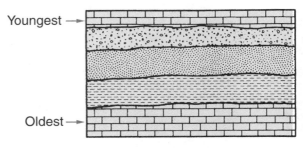

Figure 3.2-1. In a stack of sedimentary rocks, the oldest layer is at the bottom and the youngest layer is at the top.

- Sedimentary rock layers can be matched by the fossils found in them. As you learned in the previous lesson, rock layers containing the same fossils were deposited at the same time, even if the rocks are different. (Look back at Figure 3.1-2 on page 58.)

Rocks and Geologic Events

Geologists have pieced together much of Earth's history by studying rocks all over the world. Rocks contain clues that can reveal the geologic history of an area. By geologic history we mean the sequence of geologic events and processes that changed Earth's surface. For example, sedimentary rocks indicate that an area was once covered by water. Fossils in sedimentary rocks offer clues that may indicate the type of environment that existed when the organism from which the fossil formed was alive. They also provide information about changes that occurred in plants and animals over millions of years. *Stop and Think:* What do igneous rocks indicate about the geologic history of an area? *Answer:* Igneous rocks indicate the presence of volcanic activity in the past.

The Relative Age of Rocks

Geologists can interpret clues in rock structure that reveal the order in which the rocks were formed.

Figure 3.2-2 (a–d) shows rock diagrams called *rock cross sections.* In Figure 3.2-2(a), the igneous rock *cuts across* sedimentary rock layers. For this to happen, the sedimentary layers must be in place first. Therefore, the sedimentary rock is older than the igneous rock.

In Figure 3.2-2 (b), the large crack or fault (faults are associated with earthquakes and will be discussed in Chapter 4) has shifted the sedimentary rock layers so that they do not match across the fault. This means that the sedimentary rock layers formed first and are older than the fault.

Figure 3.2-2 (c) is more complex. The fault has shifted the sedimentary layers, so the sedimentary rocks were formed before the fault. The igneous rock cuts across the sedimentary rock layers, so the sedimentary rock layers were also formed before the igneous rock. But the fault has *not* shifted the igneous rock. The igneous rock must have formed *after* the fault. The relative order of events forming this cross section is 1) the formation of sedimentary rock, 2) the

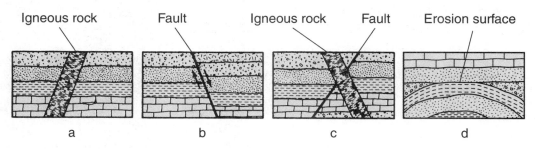

Figure 3.2-2 (a-d). Four geologic cross sections showing different examples of relative dating methods.

fault was produced, and 3) the igneous rock cuts across the sedimentary rock layers and the fault.

Figure 3.2-2 (d) is a geologic cross section that shows an erosion surface. The sedimentary rock layers were bent upward and then eroded to produce a flat land surface. New sedimentary rock was deposited on top of the eroded rock surface. An erosion surface indicates that the land was uplifted, that is, lifted up. The uplift and erosion of land is often associated with mountain building.

SKILL EXERCISE—*Determining the Sequence of Events*

Many events from Earth's past are recorded in rocks. By examining rock features, such as bends (folds), large cracks (faults), igneous rock intrusions (magma injected into pre-existing rock), and eroded surfaces, the sequence of events that produced present-day rock structures can often be determined.

Examine the diagram carefully and answer the questions that follow.

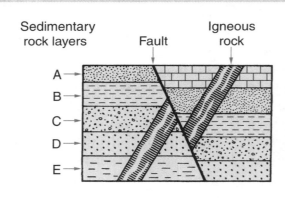

Questions

1. Which rock layer formed first? How do you know?

2. Under what conditions do sedimentary rocks usually form?

3. Which event occurred second: faulting or igneous rock intrusion? How do you know?

4. List in order the three geologic events indicated by this rock diagram.

Activities

1. Using clay (or *play dough*), construct a model that represents a cross section of a rock structure. You need three different colors of clay for this model. Follow these directions to construct your model:

(a) Flatten the clay into sheets (each a different color) about 10 cm x 10 cm x 0.5 cm.

(b) Place the sheets on top of each other so that they appear as layered sedimentary rocks.

(c) Fold the three layers upward.

(d) Carefully cut the top 1 cm off the upfold (the layers folded upward).

(e) Place another flat layer (any color) 3 cm x 10 cm x 0.5 cm on top of the cut surface.

2. Draw a diagram of the cross section model you produced.

3. List in order the four geologic processes your model represents.

Questions

1. Valerie identified the bedrock exposed by a road cut near her school as shale, a sedimentary rock. What does this suggest about the region's past?
   (1) It was once the site of underground volcanic activity.
   (2) It was once the site of surface volcanic activity.
   (3) At one time, the area was probably underwater.
   (4) Great heat or pressure once affected the area.

2. By studying folds, faults, volcanic intrusions, and erosion surfaces in rocks, geologists can determine the
   (1) absolute age of the rocks
   (2) order of past Earth events
   (3) depth of the ocean
   (4) type of climate that existed

3. The diagram shows layers of sediments deposited in a body of water. Which layer was deposited first?
   (1) layer *A*
   (2) layer *B*
   (3) layer *C*
   (4) layer *D*

4. An erosion surface in a rock structure represents what type of geologic event?
   (1) uplift and mountain building
   (2) uplift and emerging sea
   (3) volcanic activity
   (4) deposition of sediment

5. Which principle of geology tells us earthquakes occurred million of years ago?
   (1) The oldest rock layer is on the bottom.
   (2) An unconformity is a buried erosion surface.
   (3) The same geologic processes occurring today occurred in the past.
   (4) The past is the key to the future.

6. A rock cross section can be used to determine
   (1) the absolute age of rocks
   (2) the relative age of rocks
   (3) how rocks will change in the future
   (4) how rocks formed

7. Identifying an igneous rock in a cross section indicates the area was affected by
   (1) uplift and erosion
   (2) volcanic activity
   (3) earthquake activity
   (4) water

Water — Settling of particles

Top A
     B   Layers of
     C   sediments
Bottom D

Thinking and Analyzing

Base your answers to the questions on the diagram to the right and on your knowledge of science.

(a) Which rock type is the oldest?

(b) Explain how rock G formed.

(c) Explain what happened after rock G formed.

(d) Explain what happened after rock A formed.

(e) Place the following events in order from oldest (1) to youngest (4):

_____ uplift and erosion

_____ volcanic activity

_____ formation of conglomerate

_____ formation of limestone

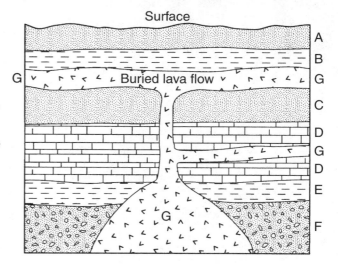

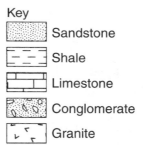

Key

Sandstone

Shale

Limestone

Conglomerate

Granite

3.3 How Do You Determine the Absolute Age of Rocks?

Objectives

Explain what is meant by the half-life of a substance.

Explain how geologists determine the absolute age of rocks.

Terms

radioactive dating: a method to determine the age of a rock by measuring its radioactivity

half-life: the amount of time it takes for half the atoms of a radioactive substance to change into nonradioactive atoms.

Absolute Dating Method

The relationship between rock layers and geologic processes can indicate the relative order of events in Earth's past. (See Lesson 3.2.) However, it cannot reveal the absolute age of the rocks (how long ago they formed). To determine the age of rocks, geologists use a technique called **radioactive dating**. This technique determines the age of a rock by determining the percentage of radioactive atoms in the rock.

Many rocks contain radioactive elements. A radioactive substance is an unstable atom that emits radiation. As it emits radiation, the atom decays and changes into a stable atom, called the decay product. The decay product is *not* radioactive. This decaying occurs at a predictable rate. This rate is known as its half-life. **Half-life** is the amount of time it takes for half the atoms of a radioactive

substance to change into nonradioactive atoms. (Nonradioactive atoms are atoms that are *not* radioactive.) Table 3.3-1 lists the half-life and decay product of some radioactive elements.

Table 3.3-1. Half-Lives of Some Radioactive Elements

Radioactive Element	Decay Product (Nonradioactive Element)	Half-Life
Uranium-238	Lead-206	4.5 billion years
Potassium-40	Argon-40	1.3 billion years
Uranium-235	Lead-207	713 million years
Thorium-232	Lead-208	14.1 billion years
Carbon-14	Nitrogen-14	5700 years

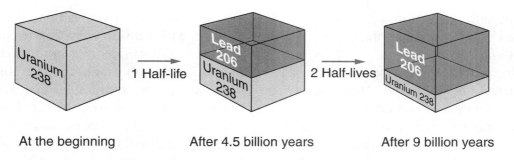

At the beginning   After 4.5 billion years   After 9 billion years

Figure 3.3-1. During each half-life period of 4.5 billion years, half of uranium-238 atoms in a rock sample change (decay) into lead-206 atoms.

For example, radioactive uranium-238 has a half-life of 4.5 billion years. That means half the uranium-238 will change into lead-206 in 4.5 billion years. Half of the remaining uranium-238 will change into lead in another 4.5 billion years. So, in 9 billion years, one-fourth of the uranium-238 will remain and three-fourths will become lead-206. By measuring the amount of uranium-238 remaining in a rock, the age of the rock can be determined. (See Figure 3.3-1.)

By using radioactive dating techniques, geologists are able to assign dates to rocks that relate directly to major events in Earth's history. Mountain building, the forming of new oceans, and the first time various life forms appeared—all these major events are recorded in rock formations. Therefore, an absolute date can be assigned to each one. Using radioactive dating techniques, geologists estimate that Earth is about 4.5 billion years old.

Determining Absolute Age

Figure 3.3-2 represents an element with a half-life of 2 years. The clear squares

Radioactive Decay

| Original radioactive element: | 100% | 50% | 25% | 12.5% | 6.25% | 3.125% |
| Decay product: | 0% | 50% | 75% | 87.5% | 93.75% | 96.875% |

Half-lives	0	1	2	3	4	5
Percent of radioactive atoms	100%	50%	25%	12.5%	6.25%	3.125%
Age (in years)	0	2	4	6	8	10

Figure 3.3-2. A chart showing how the age of a radioactive substance with a half-life of 2 years can be calculated.

represent radioactive atoms, and the darkened squares represent the decay product of those atoms.

In the original radioactive substance, all of the atoms are radioactive. After 2 years (1 half-life), 8 atoms are radioactive and 8 atoms are nonradioactive, decayed products. That is, 1/2 of the atoms are radioactive. After 4 years (2 half-lives), 1/4 of the atoms are radioactive. After 6 years (3 half-lives), 1/8 of the atoms are radioactive. And after 8 years (4 half lives), 1/16 of the atoms are radioactive. So if you know the percentage of radioactive atoms in a substance, you can determine the age of the substance.

Geologists who study radioactive dating determine the number of radioactive atoms and the number of decayed product atoms in a rock. Using this ratio, they are able to determine the age of the substance. For example, if a radioactive substance has a half-life of 1.3 million years, and 1/4 of its atoms are still radioactive, it has passed through 2 half-lives and is 2.6 million years old. *Stop and Think:* In this same substance, if only 1/8 of its atoms were still radioactive, how old would it be? *Answer:* If only 1/8 of its atoms were still radioactive, it would have passed through 3 half-lives and would be 3.9 million years old.

Activity

This activity demonstrates the half-life process. Place 64 pennies, heads up, in a shoe box. Close the shoe box and vigorously shake it once. Each shake represents 10 years. Open the box. The heads-up pennies represent radioactive atoms, and the tails-up pennies represent the decay product. In a table similar to the one below, record the number of heads-up pennies in the shoe box. Remove all the tails-up pennies and place them on the table. Record the total number of tails-up pennies on the table. Cover and shake the shoe box again and record your results again. Continue this procedure until all the pennies are removed from the box.

Shakes	0	1	2	3	4	5	6	7	8
Number of years	0	10	20	30	40	50	60	70	80
Heads-up in box	64								
Tails-up on table	0								

(a) How many times did you have to shake the shoe box until you had removed all of the pennies from it?

(b) Statistically, you should have to shake the box seven or eight times before you could remove all the pennies. Compare this with the actual number of times you shook the box.

(c) This activity represents a model of the half-life process. What is the half-life of the pennies?

Interesting Facts About Radioactive Dating

Carbon-14 (also called *radiocarbon*) has a half-life of about 5700 years. Living things take in carbon-14 during their lives and maintain a constant level of it throughout their lives. When they die, the carbon-14 decays at a constant rate known as its half-life. Therefore, a dead organism's age can be determined by measuring the amount of carbon-14 remaining. Wood tools and utensils, paper products, and bones can all be dated accurately using the carbon-14 method. Carbon-14 dating is accurate for objects up to about 50,000 years in age.

The following story demonstrates the use of carbon-14 dating.

In 1991, two tourists in the Alps discovered a human body embedded in a glacier. Scientists named the body Otzi the Iceman. At first, it was believed that Otzi had recently fallen into the glacier. Perhaps, it was thought, he fell into a large crack in the ice while rock climbing or hiking on the glacier. After extensive examination, it was determined that Otzi was between 30 and 45 years old and about 5 feet 3 inches (1.6 m) tall. Using radioactive dating of carbon14, scientists further determined that Otzi died about 5300 years earlier, in the year 3300 B.C.

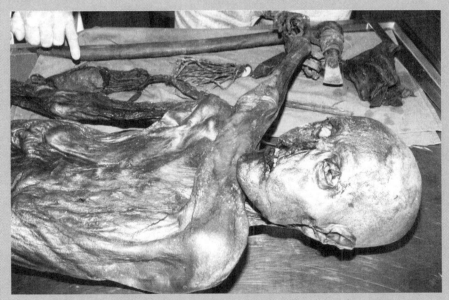

Otzi

Questions

1. Geologists use radioactive dating to
   (1) determine the absolute age of a rock
   (2) determine the order of events in Earth's geologic history
   (3) identify the rock type in a region
   (4) learn about past climate conditions

2. What is the ratio of radioactive atoms to atoms of decay product after three half-lives?
   (1) 1:2   (3) 1:4
   (2) 1:3   (4) 1:7

3. How much is 1/2 of 1/2?
   (1) 1/4   (3) 1/2
   (2) 1/8   (4) 2

4. A piece of wood was found buried in the glacial sand in Michigan. The glaciers buried the wood about 20,000 years ago. Which radioactive substance should be used to determine the age of the wood?
   (1) uranium-238, with a half-life of 4.5 billion years
   (2) uranium-235, with a half-life of 713 million years
   (3) potassium-40, with a half-life of 1.3 billion years
   (4) carbon-14, with a half-life of 5700 years

5. To determine the absolute age of a rock, the rock must
   (1) contain fossils
   (2) contain radioactive elements
   (3) form underground
   (4) form on Earth's surface

6. The rate that a radioactive element changes into a non-radioactive element is called
   (1) dating
   (2) decay rate
   (3) whole-life
   (4) half-life

7. Over a period of time a radioactive element in a rock
   (1) increases and the decay product increases
   (2) increases and the decay product decreases
   (3) decreases and the decay product decreases
   (4) decreases and the decay product increases

Thinking and Analyzing

1. Base your answers to the questions on the diagram below and on your knowledge of science. Diagram A shows a pile of 32 radioactive atoms (light circles) with a half-life of 10 minutes. Diagram B shows that after 10 minutes, half the atoms have decayed (dark circles).
(a) Predict how the atoms will change after 20 minutes. Draw a diagram showing this change.
(b) How many decayed atoms will exist after 30 minutes?
(c) How long will it take for only 2 radioactive atoms to remain?

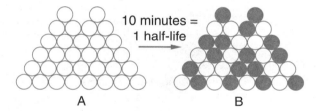

A   B

2. Potassium-40 decays into argon-40 with a half-life of 1.3 billion years. A rock sample contains 50 percent of the original radioactive potassium-40 content it had.
(a) How old is the rock?
(b) Explain why potassium-40 would not be used to date lava from a recent volcanic eruption.

3. Read "Interesting Facts About Radioactive Dating" and answer the following questions.
(a) What is another name of carbon-14?
(b) Describe how the amount of carbon-14 changes when an organism dies.
(c) Carbon-14 has a half-life of about 5700 years. Otzi the Iceman's remains are about 5300 years old. About what percent of carbon-14 remains in Otzi's body?

3.4 What Is the Geologic Time Scale?

Objectives

Describe the geologic time scale.

Describe the major divisions of the geologic time scale.

Terms

geologic time scale: a detailed chart that shows the geologic and biologic history of Earth

Precambrian (pree-KAM-bree-un) **era:** period of time on Earth (first 4 billion years) for which the fossil and rock records are poorly preserved

Paleozoic (pay-lee-uh-ZOH-ik) **era:** period of time on Earth (251–544 million years ago) when early life forms existed; animals without backbones, fish, and amphibians dominated

Mesozoic (mez-uh-ZOH-ik) **era:** period of time on Earth (65–251 million years ago) when "middle life" forms existed; reptiles (dinosaurs) dominated

Cenozoic (sen-uh-ZOH-ik) **era:** period of time on Earth (present–65 million years ago) when recent life forms existed; mammals dominated

History of Earth

Scientists believe the universe began between 13 and 15 billion years ago. About 4.5 billion years ago, Earth started forming around the sun. The age of Earth (4.5 billion years) is inferred from the age of the oldest rocks on Earth (about 4 billion years); the age of moon rocks (4.6 billion years); and the age of meteorites (4.6 billion years). They are all believed to have formed at about the same time.

Little is known about Earth during its first 4 billion years. Most of the early rock record was destroyed by weathering and erosion. The fossil record from this time shows that only simple life forms existed. About 544 million years ago, life forms on Earth started to flourish. Since then, changes in both land and life were recorded in the rocks and show how Earth evolved into the planet we are familiar with today.

Earth's Time Line

A *time line* can be drawn to represent the history of Earth. (See Figure 3.4-1.) The beginning of the line represents the origin of Earth and is labeled "4.5 bya" (billion years ago). The end of the line represents the present, or today, and is labeled "0 bya." Along the time line, labels show when

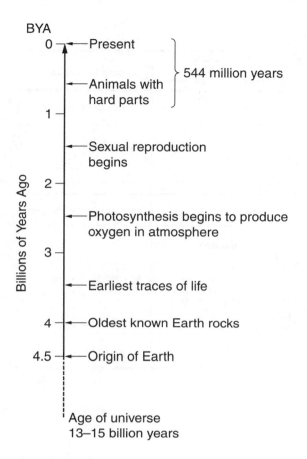

BYA

- Present
- 544 million years
- Animals with hard parts
- Sexual reproduction begins
- Photosynthesis begins to produce oxygen in atmosphere
- Earliest traces of life
- Oldest known Earth rocks
- Origin of Earth

Billions of Years Ago

Age of universe
13–15 billion years

Figure 3.4-1. A simple time line showing some geologic and biological events affecting Earth.

major events occurred on Earth. As you can see in the time line, the last 544 million years is a small portion of Earth's history.

The Geologic Time Scale

The **geologic time scale** is a detailed chart that shows the geologic and biological history of Earth. (See Figure 3.4-2 on page 76.) It expands our simple time line above by adding relative age and absolute age to the many events that created our present day Earth.

The geologic time scale is divided into four long periods of time called *eras*. The **Precambrian era** is the longest. As mentioned before, little is known about the Precambrian because the fossil and rock record is poorly preserved. Much more is known about the other three eras. The **Paleozoic era** (means "early life") started about 544 million years ago and ended about 251 million years ago. Early life forms such as animals without backbones, fish, and amphibians dominated. The **Mesozoic era** (means "middle life") started about 251 million years ago and ended about 65 million years ago. Reptiles, especially dinosaurs, dominated animal life during this era. The **Cenozoic era** (means "recent life") started about 65 million years ago and continues today. Mammals dominate animal life throughout the Cenozoic era.

Between each of the eras, there is a boundary represented by a dark line, as shown in Figure 3.4-2. The boundaries are associated with widespread geologic events, such as mountain building, earthquakes, and volcanic activity. A meteor impact is also linked with the Mesozoic-Cenozoic boundary. These events drastically affected the environmental living conditions on Earth. These events also caused the extinction of many species while, at the same time, new life forms began to appear and flourish.

The Paleozoic, Mesozoic, and Cenozoic eras are further divided into eleven *periods* of time. These periods are based on specific rock formations that were produced during each period. Each geologic period is named for the region in which these particular rocks were first found and studied. For example, the Devonian period was named

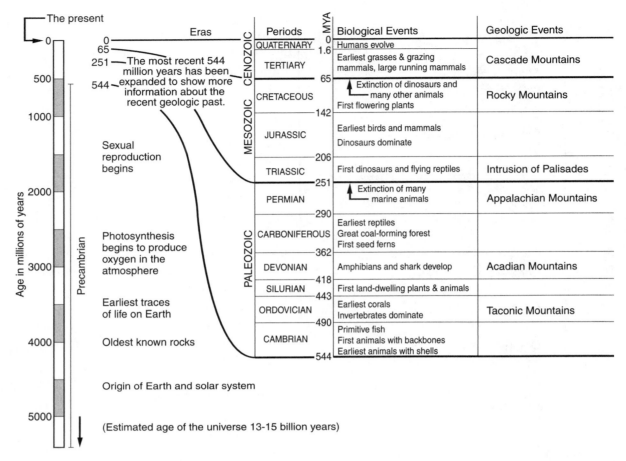

Figure 3.4-2. The geologic time scale showing major divisions of time and related geologic and biological events.

for the rocks produced between 418 and 362 million years ago that were first found and studied in Devonshire, England. (See Figure 3.4-2.)

Time periods are associated with geologic and biologic events. The events are listed in Figure 3.4-2. Geologic events are connected to the formation of mountains, and the biologic events show the evolution, or slow change, in life forms. The geologic time scale shows how organisms evolved from relatively simple forms to more complex forms as the rocks become younger.

SKILL EXERCISE—*Placing Geologic and Biological Events in Order of Occurrence*

*T*he list of geologic and biological events below is out of order. Write the list in the correct order in your notebook. Place the oldest event at the bottom of the list and the youngest (most recent) event at the top of the list. Next to each event, state whether it is a geologic or biological event.

- Humans evolved on Earth
- Extinction of the dinosaurs
- Organisms with hard parts developed
- Rock deposits produced during the Devonian period
- Oldest known rocks on Earth
- Volcanic intrusions produced the Palisades
- Large mammals appear on Earth
- Appalachian Mountains formed

Activity

In this activity, you will draw a time line that represents your life. The activity assumes you are 12 years old.

- Draw a 24-cm-long horizontal time line to represent your life.
- Divide the line into 12 equal units. Each unit represents 1 year of your life.
- Place an arrow on the right side of the line and label it "0 yrs—present."
- Label the left side of the line "12 yrs ago—birth."
- Select 10 events in your life and for each event place a point on the line. Label each point. (Examples: started school, a trip you took, met a new friend, started walking, read a book, etc.)
- Finally, divide the line into three eras: pre-child, paleo-child, and meso-child.

In which era do you remember the most events? In which era do you remember the fewest events?

Interesting Facts About a Mass Extinction

The fossil record indicates that there have been at least five mass extinctions on Earth. A *mass extinction* occurs when large numbers of species die out suddenly. Species become extinct when they do not adapt to environmental changes, or when they can no longer compete with other organisms. It is estimated that more than 99 percent of all species that ever lived on Earth have become extinct.

A large mass extinction occurred at the Cretaceous (kri-TAY-shus)-Tertiary (TUR-she-ar-ee) boundary (65 mya). (See Figure 3.4-2.) The extinction caused about 70 percent of all species to die out, including the last of the dinosaurs.

Some of the many theories presented for this mass extinction include

- Global temperature change
- Organisms died from disease
- Volcanoes covered Earth with ash
- A nearby star exploded
- An asteroid hit Earth

The asteroid theory was proposed in 1980. The theory appeared to be a possibility, but lacked substantial evidence to back it up. However, in 1990, a crater was discovered in the Gulf of Mexico near the Yucatan Peninsula. The asteroid that produced the crater was 10 km (6 miles) wide. It hit Earth about 65 million years ago. Then, in 1997, seafloor sediments collected east of Florida revealed a layer of ash that had been produced by the impact of the asteroid. The evidence of the crater and the widespread ash support the theory that a major extinction was caused by an asteroid that hit Earth 65 million years ago.

Questions

1. The geologic time scale is divided into four large time units called eras. These time divisions are based upon
   (1) the rise and fall of the ocean
   (2) locations where rocks are found
   (3) equal units of geologic time
   (4) major changes in life forms

2. The age of Earth is about
   (1) 13 billion years
   (2) 544 million years
   (3) 4.5 billion years
   (4) 251 million years

3. Geologic eras are divided into smaller time units called periods. What are these divisions based upon?
   (1) geologic events only
   (2) biological events only
   (3) geologic and biological events
   (4) equal units of geologic time

4. In which era of geologic time are humans living?
   (1) Cenozoic   (3) Paleozoic
   (2) Mesozoic   (4) Precambrian

5. Which era spans the longest period of time?
   (1) Cenozoic   (3) Paleozoic
   (2) Mesozoic   (4) Precambrian

6. Which organisms have been on Earth the longest period of time?
   (1) humans   (3) fish
   (2) dinosaurs   (4) birds

Thinking and Analyzing

1. Base your answers to the following questions on the diagram at the right and on your knowledge of science. The diagram shows a portion of the geologic time scale.
   (a) How long did the Paleozoic era last?
   (b) Explain why it is rare to find a fossil greater than 544 million years old.
   (c) Explain what is meant by the statement, "A species becomes extinct when it *does not* adapt to environmental changes."

2. Use the geologic time scale on page 76 to help answer the following questions.
   (a) What geologic event occurred during the Permian period?
   (b) The Catskill Mountains in New York State contain Devonian rocks. You cannot find fossils of dinosaurs in the Catskill Mountains. Why?
   (c) Coal beds are found in the rocks of Pennsylvania. During what period of time did the coal beds form?

Portion of the geologic time scale

Era	Period	Period started (MYA)
Cenozoic	Quaternary	1.6
	Tertiary	65
Mesozoic	Cretaceous	142
	Jurassic	206
	Triassic	251
Paleozoic	Permian	290
	Carboniferous	362
	Devonian	418
	Silurian	443
	Ordovician	490
	Cambrian	544

MYA = Million of years ago

Review Questions

Term Identification

Each question below shows two terms from Chapter 3. One of the terms is defined. (1) Choose the term that matches the definition. (2) Describe how the two terms are different. Following each term is the section (in parenthesis) where the description or definition of the term is found.

1. *Mold (3.1) — Cast (3.1)*
   A hollow area in sedimentary rock produced when a fossil dissolves out of the rock

2. *Original remains (3.1) — Replaced remains (3.1)*
   A frozen woolly mammoth from Siberia is an example of this

3. *Relative dating (3.2) — Absolute dating (3.3)*
   Age determined using radioactive elements

4. *Uranium-238 (3.3) — Lead-206 (3.3)*
   A decay product that is not radioactive

5. *Potassium-40 (3.3) — Carbon-14 (3.3)*
   Used to date organic material that is 50,000 years old or younger

6. *Geologic time scale (3.4) — Time line (3.4)*
   A chart that shows the geologic and biological history of Earth

7. *Paleozoic (3.4) — Cenozoic (3.4)*
   The era that exists today and is referred to as the "recent stage of life"

8. *Era (3.4) — Period (3.4)*
   Largest interval of geologic time associated with stages of living organisms

Multiple Choice (Part 1)

Choose the response that best completes the sentence or answers the question.

1. The remains or traces of organisms that lived long ago are called
   (1) evolution
   (2) fossils
   (3) geologic cross sections
   (4) relative dating

2. Fossils became abundant 544 million years ago because
   (1) life began
   (2) organisms with hard parts became abundant
   (3) rocks started forming
   (4) Earth formed

3. What do fossils tell us about the history of Earth?
   (1) how rocks formed
   (2) what environmental conditions existed
   (3) the absolute age of the rocks
   (4) type of geologic processes that occurred

4. Which method is most reliable in matching rocks from two different locations?
   (1) identifying physical characteristics of the rocks
   (2) identifying the mineral content of the rocks
   (3) identifying similar fossils in the rock layers
   (4) identifying the thickness of the rock layers

5. Which item provides evidence that life forms and the environment have both changed over the past 544 million years?
   (1) volcanoes
   (2) earthquakes
   (3) fossils
   (4) erosional features

6. Dinosaur footprints are an example of what type of fossil-producing process?
   (1) original remains
   (2) replacement
   (3) mold and cast
   (4) impression

7. The diagrams below show stages in the development of a landscape. In which stage is erosion the most dominant force?
   (1) stage 1   (3) stage 3
   (2) stage 2   (4) stage 4

8. Extinction of a species is most likely to occur as a result of
   (1) evolution
   (2) migration
   (3) selective breeding
   (4) environmental changes

9. In which type of rock are fossils usually found?
   (1) igneous   (3) sedimentary
   (2) metamorphic   (4) volcanic

10. The geologic cross section below shows fossils and the rock layers in which they are found. Crustal movement has not displaced the rock layers.

(Not drawn to scale)

Which fossil is considered the oldest in the cross section shown?

   (1) armored fish   (3) early horses
   (2) dinosaurs   (4) trilobites

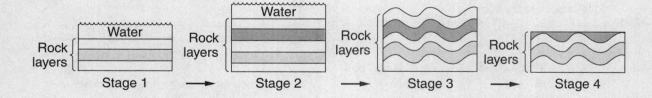

11. An erosion surface in a geologic cross section indicates that the land was at some time
  (1) below sea level
  (2) above sea level
  (3) covered by lava flows
  (4) deep underground

12. From which era of Earth's history are fossils rarely found?
  (1) Cenozoic
  (2) Mesozoic
  (3) Paleozoic
  (4) Precambrian

Thinking and Analyzing (Part 2)

1. Base your answers to the questions on the diagram below and on your knowledge of science. A rock was found containing fossils of three different ancient species: cephalopods, gastropods, and brachiopods. The table shows the time ranges during which these organisms lived.
  (a) About how many years did the brachiopods live?

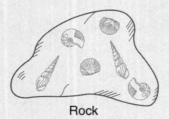

Rock

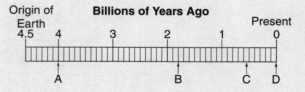

(b) The rock must have been formed at a time when all three species were living. What is the best estimate of when the rock was formed?
(c) What happened to the brachiopods 18 million years ago?

2. Base your answers to the following questions on the table below and on your knowledge of science. The table shows how a radioactive substance and its decay product change during the first three half-lives.

Number of half-life	Start	1	2	3
Radioactive substance (%)	100	50	25	12.5
Decay product (%)	0	50	75	87.5

(a) Carbon-14 has a half-life of 5700 years. If a piece of wood contains 25% carbon-14 (radioactive substance) and 75% nitrogen-14 (decay product), how many half-lives has the wood aged?
(b) What is the age of the wood?
(c) What would be the percent of radioactive carbon-14 and the percent of its decay product nitrogen-14 after 4 half-lives?

3. Base your answers to the question on the diagram below and on your knowledge of science. The time line below represents the 4.5 billion-year history of Earth.

(a) Which letter best shows the time humans inhabited Earth?
(b) Which letter best shows the end of the Precambrian era?
(c) What era is referred to as "middle life"?

4. Base your answers to this question on the reading passage and diagram below and on your knowledge of science.

Trilobite

The fossil in the picture at the bottom of the page is a trilobite. Trilobites were crablike animals that lived at the bottom of the ocean for 300 million years. Most were less than 10 cm in length, but some were as long as 45 cm. There were probably thousands of different kinds of trilobites. Trilobites had a hard external skeleton and are among the oldest fossils.

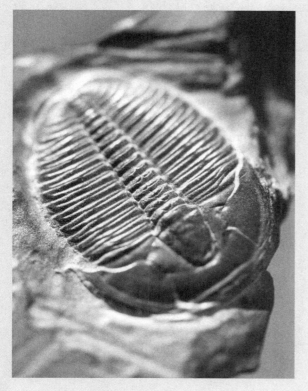

(a) The trilobite in the picture is a cast. Explain how a cast forms.
(b) In what type of rock might you find a trilobite fossil?
(c) During which era did trilobites live?

5. Base your answers to this question on the reading passage and diagram below and on your knowledge of science.

Preserved in Amber

Sap is a substance secreted by some trees. Many years ago, plants and small animals were caught in the sap on the trees. Sap hardens and turns into a clear substance called amber. The plants or animals are preserved as fossils in the amber. Part of a plant preserved in amber is shown below.

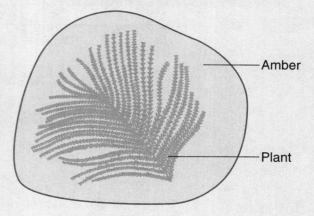

(a) Explain why fossils are important to scientists.
(b) Explain why plant fossils found in amber, such as the one shown, are *not* usually found in volcanic rocks.

Chapter Puzzle (*Hint:* The words in this puzzle are terms used in the chapter.)

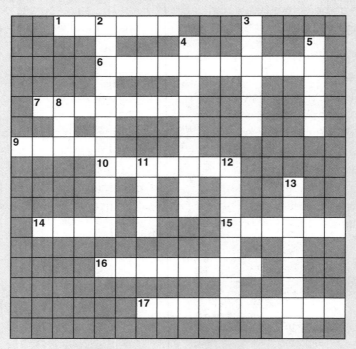

Across

1 used to date 3000-year-old mummy

6 longest era

7 sequential order determination of rock age

9 geologic time _____ shows geologic and biological history of Earth

10 elephant-like fossil found in Siberia

14 filled mold

15 remains or trace of organisms that lived long ago

16 recent life era

17 dinosaur footprint-type fossil

Down

2 process producing petrified wood

3 era time division

4 era of early life

5 type of fossil used to match rock layers

8 longest division of geologic time

11 hollow depression produced by a fossil

12 time it takes for a radioactive substance to decay

13 dinosaur era

Chapter 4
Plate Tectonics

Contents

The explosive eruption of Mount St. Helens on May 18, 1980.

What Is This Chapter About?

The theory of continental drift was proposed almost 100 years ago. At first, the scientific community did not accept the theory. As additional evidence was discovered, the theory of plate tectonics was born. The theory of plate tectonics provides a better understanding of earthquakes, volcanoes, and mountain ranges.

In this chapter you will learn:

1. There is evidence that supports the theory of continental drift.

2. Earth's crust is broken into pieces called plates. Convectional heat from Earth's interior causes the plates to move.

3. Most earthquakes, volcanoes, and mountains are located along plate boundaries.

4. Topographic maps show the shape and form of land features such as mountains, valleys, and lakes.

Science in Everyday Life

Most earthquakes in the United States occur in Alaska and California. A serious earthquake can occur anywhere, however. A major earthquake struck Charleston, South Carolina in 1886. Four major earthquakes struck Missouri between December 1811 and February 1812. Although your risk may be low, you should be aware of the dangers associated with earthquakes.

Internet Sites:

http://www.ucmp.berkeley.edu/geology/tectonics. html Animated graphics show the motion of the continents during the last 750 million years. Information about the geologic time scale, the history of plate tectonics, and the tree of life is also presented.

http://www.earthquake.usgs.gov/regional/neic/ The National Earthquake Information Center provides information about past and present earthquakes. It maintains a current worldwide earthquake list.

http://www.volcano.und.edu/ Volcano World has information about past and present volcanic eruptions. Many outstanding pictures are available.

85

4.1 What Is Continental Drift?

Objectives

Describe the evidence that supports Alfred Wegener's theory of continental drift.

Explain why the theory of continental drift was not accepted at first.

Term

theory of continental drift: a theory that suggests that the continents were once a single mass of land that broke apart and are currently drifting across Earth's surface

The Theory of Continental Drift

Look at a map of the world. Do the continents on each side of the Atlantic Ocean look like they fit together? Although it is not a perfect fit, they do look like they fit like pieces in a jigsaw puzzle. In 1912, Alfred Wegener (1880–1930), a German meteorologist, proposed the **theory of continental drift**. The theory suggests that the continents were once a single mass of land that broke apart and are drifting across Earth's surface. (See Figure 4.1-1.) It also suggests that the Atlantic Ocean was formed by the separation of North America and South America from Europe and Africa.

Evidence of Continental Drift

Alfred Wegener thought that the continents were once joined in a single landmass that broke apart about 200 million years ago. These newly created continents slowly drifted apart over a period of millions of

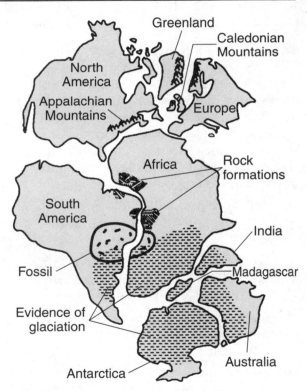

Figure 4.1-1. Continental shapes fit together like pieces of a puzzle. Fossils, rock formations, mountains, and glacial features match.

years. In addition to the fact that the shapes of the continents matched like a jigsaw puzzle, Alfred Wegener gathered other evidence for his theory. He showed how fossils, mountain ranges, rock formations, and glacial features were also very much alike on continents that were separated by the Atlantic Ocean. The following sections review the evidence Alfred Wegener gathered to support his theory.

Fossils

Identical fossils have been found in South America and Africa. Fossils of a specific type of fern and an aquatic reptile found on two landmasses so far apart were difficult to explain. Alfred Wegener suggested that these organisms once lived on a single landmass that split and the pieces drifted apart.

Mountain Ranges

When pieces of a jigsaw puzzle are put together, the shape of the pieces must match, but the features of the picture must match, too. When the continents are brought together, you expect land features to match. They do! The Appalachian Mountains in eastern North America match the mountains in Great Britain, Greenland, and Europe. When they are "put together," they form a continuous chain of mountains.

Rock Formations

Alfred Wegener also identified similar rock formations on both sides of the Atlantic. Some rocks along the east coast of South America match rocks along the west coast of Africa.

Glacial Features

Alfred Wegener recognized that glacial features found in South America, Africa, India, Australia, and Antarctica were formed at the same time. As glaciers move, they deposit sand and scratch underlying rock. The distribution of sand and scratches on these continents indicate they were connected at the time the glaciers were present.

The scientific community discussed continental drift for about two decades, but eventually interest in the theory declined. Many scientists argued that the cause of continental drift was not explained. They argued that although the evidence presented by Alfred Wegener was intriguing, his theory could not identify the force causing the continents to move. It took many scientific studies by many scientists over the next 50 years to discover the cause of the drifting continents.

Activities

1. Obtain a map that shows the outlines of the continents of the world. (You can go to *http://www.eduplace.com/ss/maps/.*) Cut out the continents. Place the pieces on a blank sheet of paper so they fit together to form a single landmass. Glue or tape the pieces onto the paper. Label each of the continents. Using different color crayons, identify similar areas of fossils, mountain ranges, rock formations, and glaciation.

2. Take a page out of a newspaper and tear it into eight irregularly shaped pieces. Ask several people to correctly put the pieces together. After they put the pieces together, ask them to explain how they did it. Their answers should include the matching of shapes and features (pictures and words) on the pieces of paper. Describe in your notebook how this activity relates to the theory of continental drift.

Questions

1. Early recognition of continental drift came from
   (1) observing that the land on each side of the Atlantic Ocean fits like a puzzle
   (2) identifying earthquake and volcanic belts
   (3) making measurements of the distance across the Atlantic Ocean
   (4) determining the age of rocks under the Atlantic Ocean

2. Why was the theory of continental drift rejected at first?
   (1) It was not proposed by a geologist.
   (2) The shape of the continents did not fit perfectly.
   (3) The cause of the motion could not be identified.
   (4) The rocks were too old.

3. According to Alfred Wegener's theory of continental drift, which ocean formed when North and South America separated from Europe and Africa?
   (1) Atlantic   (3) Arctic
   (2) Pacific   (4) Indian

4. In addition to the shape of the continents, what other evidence did Alfred Wegener present for his theory of continental drift?
   (1) the size of the Atlantic Ocean
   (2) the matching of volcanoes in South America and Africa
   (3) the matching of fossils in South America and Africa
   (4) the matching of mountains in South America and Africa

5. The theory of continental drift suggests that the continents started drifting apart about
   (1) 10,000 years ago
   (2) 100,000 years ago
   (3) 1 million years ago
   (4) 200 million years ago

Thinking and Analyzing

1. Base your answers to the questions on the map below and on your knowledge of science. The map shows the continents on each side of the Atlantic Ocean.

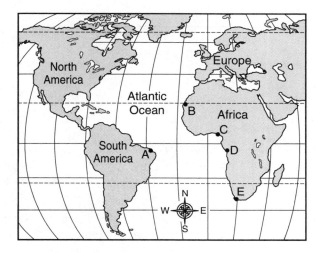

(a) According to the theory of continental drift, how is the distance between South America and Africa changing?
(b) If Precambrian granite were found at location A in South America, at what location do you expect to find Precambrian granite in Africa?

2. Base your answers to this question on the reading passage below and on your knowledge of science.

In 1929, Alfred Wegener wrote the following passage in the fourth edition of his book *The Origins of Continents and Oceans*:

> "Scientists still do not appear to understand sufficiently that all Earth sciences must contribute evidence toward unveiling the state of our planet in earlier times, and that the truth of the matter can only be reached by combing all this evidence...."

(a) Explain what Wegener meant by this statement.
(b) Describe two pieces of evidence Wegener presented for his theory of continental drift.
(c) Why wasn't Wegener's theory of continental drift accepted at first?

What Is Plate Tectonics?

Objectives

Describe how the Mid-Atlantic Ridge produces new ocean floor.

Explain what causes crustal plates to move.

Terms

mid-ocean ridge: a long chain of mountains on the ocean floor

trench: a very deep, elongated gorge with steep walls in the ocean floor

seafloor spreading: the way the seafloor moves, like a conveyor belt, away from a mid-ocean ridge

theory of plate tectonics (tehk-TAHN-ihks)**:** the theory that proposes that Earth's crust consists of moving plates

convection cell: the motion of Earth's mantle caused by the upward heat transfer inside Earth

---

Ocean Floor Features

Strong evidence for Alfred Wegener's theory of continental drift came after his death. During World War II (1941–1945), improved mapping technology revealed many features on the ocean floor. Chains of mountains, high single mountains, flat plains, and deep gorges were discovered. During the next two decades, further study of the ocean floor revealed details that improved our understanding of activity taking place in Earth's crust.

Figure 4.2-1 shows a general profile of the ocean floor. Portions of the deep ocean floor are flat plains. However, many other

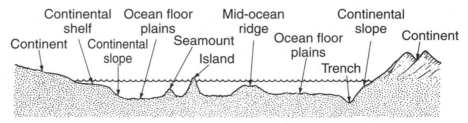

Figure 4.2-1. A general profile of the ocean floor showing some of its features.

physical features can also be found. Some ocean floor features shown in the profile are:

1. The **mid-ocean ridge** is a long chain of mountains on the ocean floor. The mid-ocean ridge is about 64,000 km (40,000 miles) long and extends around Earth. Volcanic activity occurs along the mid-ocean ridge. Islands form when volcanic activity along the ridge breaks through the surface of the ocean. Iceland and the Azore Islands in the Atlantic Ocean are formed by volcanic activity along the Mid-Atlantic Ridge.

2. **Trenches** are very deep, elongated gorges in the ocean floor. They are the deepest part of the ocean floor. The Marianas Trench (11,033 m or 36,201 ft deep) in the Pacific Ocean and the Cayman Trench (7,686 m or 25,216 ft deep) in the Caribbean are examples.

3. *Continental shelves* are areas of the seafloor that slope gently away from the coastlines of most continents. The slope angle is so gentle that if you could stand on a continental shelf, you might think you were on level ground.

4. *Continental slopes* drop away sharply from the outer edges of continental shelves to the deep ocean floor.

5. *Seamounts* are high underwater mountains that rise above the deep ocean floor. Most seamounts are volcanoes.

6. *Islands* form not only from volcanic activity along the mid-ocean ridge as mentioned before, but also form when the tops of seamounts rise above the surface of the ocean. The Bermuda islands are an example of seamounts.

Continental Drift Revisited

During the 1940s and 1950s, data collected from the ocean floor provided new evidence for the theory of continental drift. Detailed mapping of the ocean floor showed an underwater mountain ridge (the Mid-Atlantic Ridge) in the middle of the Atlantic Ocean. Radioactive dating techniques, as discussed in Chapter 3, indicated that the youngest rocks were along the ridge and the oldest rocks were farthest away. This suggested that the seafloor was moving outward from the mid-ocean ridge toward the continents.

In 1960, Harry Hess (1906–1969), a geologist from Princeton University, suggested that the mid-ocean ridge was produced by lava rising up from under the crust. The ridge consists of young, newly formed rock. This newly formed rock is being pushed continuously away from the ridge by the rising lava. The conveyor-like motion of the seafloor moving away from the mid-ocean ridge is called **seafloor spreading**. (See Figure 4.2-2.) The rock becomes progressively older as you move farther east and west of the ridge.

Scientists know that, every 200,000–300,000 years, Earth's magnetic poles reverse. That is, the north and south magnetic poles switch position. When hot molten rock hardens along the mid-ocean ridge, the iron particles, acting like tiny compass needles in the rock, become fixed,

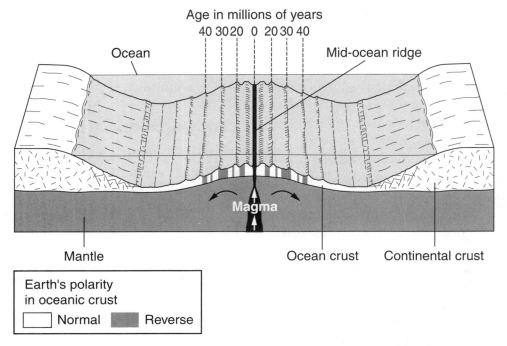

Figure 4.2-2. Supporting the theory of seafloor spreading: 1) the age of oceanic crust increases with increasing distance from the mid-ocean ridge, and 2) matching magnetic bands are found on both sides of the mid-ocean ridge.

indicating the direction of the magnetic poles at that time. In 1963, two British scientists discovered that rocks equidistant (the same distance) east and west of the ridge had matching polarity. The matching of similar polarity bands on each side of the ridge confirms that the new rock moves away from the ridge like two conveyor belts, one moving east, and the other west. (See Figure 4.2-2.) This discovery was proof of seafloor spreading.

In addition to confirming the motion of Earth's crust, magnetic bands allowed scientists to measure how fast and how far the seafloor was spreading. It is estimated that the seafloor has been moving away from the Mid-Atlantic Ridge for about 200 million years at the rate of a few centimeters per year. How fast is this? It is about as fast as your fingernails grow!

In the late 1960s, the theory of continental drift was expanded into a broad theory that brings together many geologic processes and features. It is called the *theory of plate tectonics*. The theory of plate tectonics accounts for the drifting of continents, seafloor spreading, volcanoes, earthquakes, mountains, and many other geologic processes and features. In addition, it identifies the cause of continental drift.

The Theory of Plate Tectonics

Stop and Think: Did you know that fossils of some seashells are found at the tops of mountains? How do you think

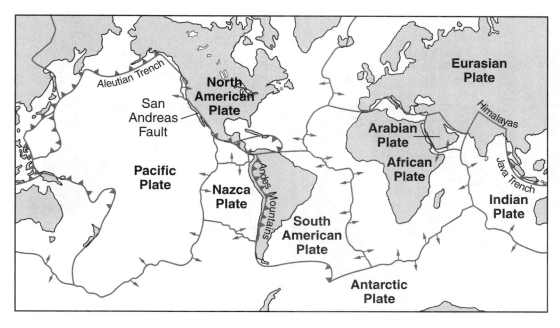

Figure 4.2-3. Earth's major crustal plates. The arrows show where plates are spreading apart. The triangular "teeth" show where one plate is sliding beneath another plate.

they got there? ***Answer:*** Shellfish live underwater. Years ago, shellfish became fossilized in sedimentary rock. Tectonic forces associated with crustal movement lifted the sedimentary rock to form mountains. The same forces today produce earthquakes and volcanic eruptions.

Geologists explain these forces and their movements with the **theory of plate tectonics**. The theory proposes that Earth's crust consists of *plates.* To help you understand this idea, think of the sections of the cracked shell around a hard-boiled egg. These sections of cracked eggshell are like the sections, or plates, on Earth's crust. (See Figure 4.2-3.) The plates on Earth's crust slowly move and interact at their boundaries, producing mountains,

volcanoes, earthquakes, and new crust.

In Chapter 1, we discussed Earth's structure and how the force of gravity pulls all matter toward the center of Earth, and that the decay of radioactive elements produces Earth's internal heat. Geologists believe Earth's internal heat is transferred upward toward the crust by convection. You may recall from previous science lessons that convection is the transfer of heat within a fluid. Earth's internal heat causes the upper mantle rock to flow like a fluid. It becomes hot and less dense, and it rises.

The transfer of heat produces large convection currents, or **convection cells**, in the upper mantle. When the rising mantle rock reaches the crust, it flows sideways,

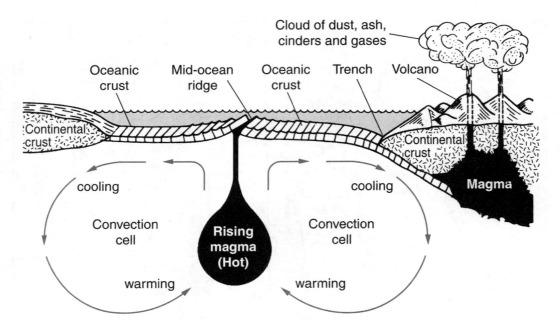

Figure 4.2-4. Convection currents in the mantle cause movements of Earth's crustal plates. This was the missing piece to Wegener's continental drift theory.

carrying with it the crustal plates. In other areas, the mantle is cooled and moves down. It is believed that the convection cells provide the force needed to move the crustal plates. (See Figure 4.2-4.)

Activity

(*Note:* The following activity should be done under the supervision of your teacher.)

Place a beaker of water over a lighted candle. Place the candle so it heats only half the bottom of the beaker. The water directly above the flame becomes hot and begins to rise. The surrounding cooler water sinks to replace the rising water. Place a few drops of food coloring in the water to observe the convection current. Now place small wooden chips on top of the rising water. What do you observe? In your notebook, draw a labeled diagram of this activity. Compare this model to the theory of continental drift.

Questions

1. The theory that Earth's crust is broken up into large pieces that move and interact is called
   (1) evolution
   (2) mountain building
   (3) the rock cycle
   (4) plate tectonics

2. Crustal plate movement is caused by
   (1) ocean currents
   (2) convection currents
   (3) the gravity of the moon
   (4) uneven heating of Earth's surface

3. Which ocean floor feature forms the deepest part of the ocean?
   (1) trench
   (2) continental shelf
   (3) mid-ocean ridge
   (4) continental slope

4. New ocean floor is produced
   (1) along colliding plate boundaries
   (2) on the edge of continents
   (3) in ocean trenches
   (4) along the mid-ocean ridges

5. The ocean floor is best described as
   (1) a flat, featureless plain
   (2) having mountains, gorges, and plains
   (3) a flat plain with a deep gorge in the center
   (4) having plains and plateaus

Thinking and Analyzing

1. Base your answers to the questions below on Figure 4.2-4 on page 94 and on your knowledge of science. Figure 4.2-4 shows the heat flow and movement of some material within Earth that causes sections of Earth's crust (plates) to move.
   (a) How does the temperature of Earth's crust compare to the temperature of Earth's interior?
   (b) Name two geologic features or events that might result from the movement of crustal plates.
   (c) How does the thickness of Earth's oceanic crust compare to the thickness of the continental crust?

2. Base your answers to the following questions on the diagram at the right and on your knowledge of science. The diagram shows a cross section view of the Mid-Atlantic Ridge.

   (a) At what letter location on the west side of the ridge does the youngest rock exist?
   (b) Describe how the position of point C will change in the future.
   (c) What number location on the east side of the mid-ocean ridge matches the magnetic polarity of letter location B on the west side of the ridge?

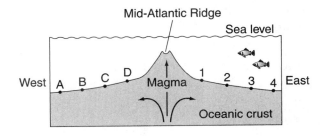

4.3 What Happens at Plate Boundaries?

Objectives

Identify the difference between continental crust and ocean floor crust.

Describe the different types of plate boundaries.

Terms

tectonic forces: forces produced usually along plate boundaries that cause rocks to bend, crack, and uplift

divergent (di-VER-gent) **boundary:** the crustal plate boundary produced when crustal plates move apart

convergent (kan-VER-gent) **boundary:** the crustal plate boundary produced when crustal plates come together

transform boundary: the crustal plate boundary produced when crustal plates slide past each another

---

Composition of Earth's Crust

Most of Earth's continental crust is composed of igneous and metamorphic rock. In some places, a relatively thin layer of sedimentary rock covers the surface.

Generally, the rock that makes up the continental crust is different from the rock that makes up the ocean floor crust. The continental crust contains rock with a chemical composition similar to granite. You may remember from Chapter 2 that granite is a light-colored igneous rock. It contains lots of quartz and feldspar. Granite also has a relatively low density that causes it to rise above the denser rock on the ocean floor. It is this rising up of granite that produces continental landmasses. (See Figure 4.3-1.)

The ocean floor crust is mainly composed of rock similar to basalt. Basalt is a dark-colored igneous rock that contains minerals rich in iron. As a result, it tends to sink under granite because it is denser.

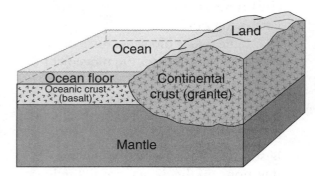

Figure 4.3-1. Continental crust is thick and has a composition similar to granite. Oceanic crust is thin and has a composition similar to basalt.

Types of Plate Boundaries

Crustal plates move slowly as a single unit across Earth's surface. Geologic activity is greatest along plate boundaries. Along some boundaries, plates are diverging, or moving apart. Along other boundaries, plates converge, or come together. At transform boundaries, plates slide past each other. The stress and strain created along plate boundaries produce **tectonic forces** that cause the crust to bend, crack, and uplift. Also, earthquakes, volcanic activity, and mountains are commonly associated with plate boundaries.

Divergent Boundary

A **divergent boundary** is produced when crustal plates move apart. As discussed in the previous lesson, the rising of hot molten rock carries the plates away from mid-ocean ridges. Thus, divergent boundaries are locations where plates are created. An example of a divergent boundary is the Mid-Atlantic Ridge. (See Figure 4.3-2A.)

Convergent Boundary

A **convergent boundary** is produced when crustal plates come together. When a high-density oceanic plate collides with a low-density continental plate, the denser oceanic plate tends to slide under the continental plate. The area where this occurs is called a *subduction zone*. (See Figure 4.3-2B.) As the oceanic plate moves downward, a trench is produced. Continental rocks are pushed up, forming mountains along coastal areas. Friction and internal heat melt the rocks at the subduction zone and produce magma. Magma, being of low density, rises and moves up toward Earth's surface, producing volcanoes along the coastlines of the subduction zone. Frequent earthquakes are also associated with subduction. The Aleutian Trench off the coast of Alaska is an example of a subduction zone.

Another type of convergent boundary is produced when two continental plates collide. (See Figure 4.3-2C.) Because continental plates have similar densities, they resist subduction. The collision of two continental plates causes massive uplift of land that produces high mountains. An

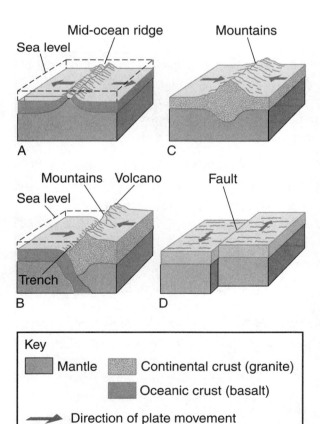

Figure 4.3-2. Plate boundaries can be classified as (A) divergent, (B and C) convergent, and (D) transform.

example of this type of boundary can be found where the Indian Plate collides with the Eurasian Plate. (Refer to Figure 4.2-3 on page 93.) The collision of these two plates created the Himalaya mountains.

Transform Boundaries

A **transform boundary** occurs when one crustal plate slides sideways past another crustal plate. (See Figure 4.3-2D.) A major break in the crust exists along a transform boundary. In California, the Pacific Plate is sliding past the North American Plate along the San Andreas Fault, producing a region of high earthquake activity.

Stop and Think: Is Washington D.C. located near any plate boundary? *Answer:* Washington D.C. is not located near a plate boundary. It is located in the center region of the North American Plate. (Refer back to Figure 4.2-3 on page 93.) The eastern edge of the plate is the Mid-Atlantic Ridge, a divergent plate boundary. The western edge of the plate is the San Andreas Fault in California, a transform boundary.

Interesting Facts About Iceland and the Mid-Atlantic Ridge

Iceland is located on top of the Mid-Atlantic Ridge. Its geology provides a unique opportunity to observe what is happening along a divergent plate boundary. New crust is created at the rate of 45 cubic km every 1000 years. The plates are moving east and west at about 1 to 2 cm per year. Some of the geologic processes and features found in Iceland are:

- *Earthquakes.* As the Mid-Atlantic Ridge splits apart, vibrations are sent through the ground. Many earthquakes occur in Iceland.

- *Volcanoes.* Iceland has more than 200 volcanoes. Many of the volcanoes are active. In 1963, geologists had an opportunity to observe a volcano on the ocean floor develop into a newly formed island south of Iceland. The island was named Surtsey.

- *Hydrothermal features.* In Iceland, magma is close to Earth's surface. The magma heats the surrounding rock and water in the ground. Heated groundwater produces hydrothermal (*hydro* means "water," and *thermal* means "heat") features. *Hot springs* are produced when heated water reaches the surface. If the heated water erupts, a hot water fountain called a *geyser* (GY-zer) is produced. Some water may be released in the form of steam called a *fumarole* (FYOO-mer-rol).

SKILL EXERCISE—*Using Math to Determine the Rate of Plate Motion*

The Hawaiian Islands are located over a volcanic "hot spot" in the middle of the Pacific Plate. A hot spot is a volcanic area in the middle of a plate where you would not expect to find volcanic activity. As the plate moves northwest over the hot spot, new islands form. The old islands erode and become seamounts. A chain of seamounts can be found in the Pacific Ocean between Hawaii and the Aleutian Trench.

The map below shows Kilauea Volcano directly above the hot spot on the eastern side of Hawaii. Kilauea has been continuously active since 1983. The islands get progressively older as you move northwest of Kilauea. Kauai, a western island, contains rocks that are 5.6 million years old. Use Figure 4.3-3 and your knowledge of math to answer the following questions.

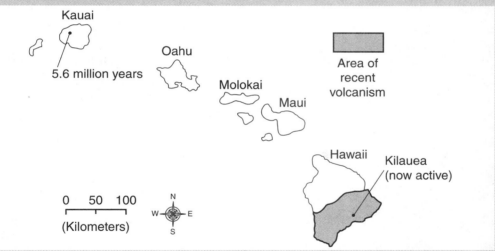

Figure 4.3-3

Questions

How many years did it take for Kauai to move from Kilauea to its present location?

How far did Kauai move in that time? (*Hint:* Use the scale to determine the distance.)

How fast is the Pacific Plate moving? _____ km/1 million years, or _____ cm/yr

In what geographic direction is the Pacific Plate moving?

Questions

1. The Mid-Atlantic Ocean is a
   (1) divergent plate boundary
   (2) convergent plate boundary
   (3) transform boundary
   (4) stationary boundary

2. What happens when an oceanic plate pushes into a continental plate?
   (1) The oceanic plate tends to move up and over the continental plate.
   (2) The oceanic plate tends to move down and under the continental plate.
   (3) Both plates tend to become stationary.
   (4) The oceanic plate tends to slide past the continental plate.

3. The Himalaya mountains were produced by the collision of
   (1) two oceanic plates
   (2) two continental plates
   (3) an oceanic plate into a continental plate
   (4) a continental plate into an oceanic plate

4. New York State is located
   (1) along a divergent plate boundary
   (2) along a convergent plate boundary
   (3) along a transform plate boundary
   (4) in the center region of a plate

5. Which statement best describes the continental crust?
   (1) It is thin and composed of a granite-like rock.
   (2) It is thick and composed of a granite-like rock.
   (3) It is thin and composted of a basalt-like rock.
   (4) It is thick and composed of a basalt-like rock.

Thinking and Analyzing

1. Base your answers to the questions on the diagram below and on your knowledge of science. The diagram shows the Mid-Atlantic Ridge located in the center of the Atlantic Ocean.

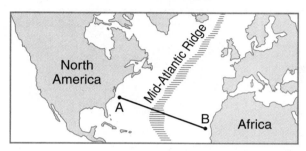

(a) How is the distance between points A and B changing?

(b) What type of plate boundary is the Mid-Atlantic Ridge?

(c) Describe two differences between continental crust and oceanic crust.

2. Base your answers to the questions on the map below and on your knowledge of science. The map shows the San Andreas Fault in California. The arrows indicate the relative motion of the Pacific Plate.

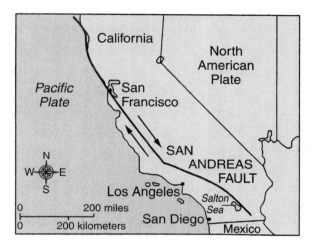

(a) In which geographic direction is the Pacific Plate moving?

(b) Along what type of plate boundary is the San Andreas Fault?

(c) Explain why there is a high risk of earthquakes in San Francisco.

4.4 What Is an Earthquake?

Objectives

Explain the cause of earthquakes.

Describe the pattern of earthquakes on Earth.

Describe how the size of an earthquake is determined.

Terms

earthquake: a sudden shaking in Earth's crust

focus (FOH-kuhs): the underground location where an earthquake occurs

epicenter (EHP-ih-sehnt-uhr): the point directly above the focus on Earth's surface

seismic (SIZE-mihk) **waves:** vibrations set off by an earthquake that travel through the earth

tsunami (tsoo-NAH-mee): a series of large ocean waves caused by an earthquake on the ocean floor

Earthquakes

An **earthquake** is a sudden shaking in Earth's crust. Moving plates rub against each other. This causes stress in the rocks that are located along plate boundaries. When the stress becomes great enough, the rocks may bend, fracture, and break, setting off vibrations that travel through the ground. The vibrations cause the ground to shake, producing an earthquake.

To help you understand this idea, bend a Popsicle stick. First it will bend. Then tiny cracks will appear in the bent area. Eventually, the stick will crack with a snapping sound. Under great pressure, rocks do the same thing.

The bending of rocks is called *folding*. When tectonic forces press against the sides of rock layers, the rock layers may bend, or fold. The land is squeezed into upfolds and downfolds, forming long mountain ridges (upfolds) and valleys (downfolds). (See Figure 4.4-1a.) *Stop and Think:* What happens when you push a loose rug against the wall? *Answer:* The rug produces a series of folds. Rocks in the crust behave in a similar manner.

Large cracks form in rock layers when tectonic forces squeeze or pull rocks beyond their ability to bend or stretch. The rocks break, producing deep cracks in the crust called *faults*. Motion along the fault causes an earthquake. (See Figure 4.4-1b.)

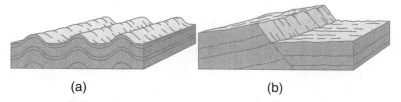

(a)   (b)

Figure 4.4-1. Forces produced by the movement of tectonic plates cause (a) folding and (b) faulting.

The underground location where an earthquake occurs is called the **focus**. The point on the Earth's surface directly above the focus is called the **epicenter**. (See Figure 4.4-2.)

When you plot many earthquake epicenters on a world map, you see that earthquakes are associated with plate boundaries, mountains, and active volcanoes. The shaded areas on Figure 4.4-3 on page 104 represent zones of high earthquake intensity. ***Stop and Think:*** How does Figure 4.4-3 compare with Figure 4.2-3? ***Answer:*** Earthquakes and plate boundaries are located in the same areas.

Measuring Earthquakes

Seismic waves are vibrations that travel through the earth. Seismic waves are detected using an instrument called a *seismograph* (SIZE-muh-graf). The waves are recorded as a zigzag line on a rotating drum. The date, time, and size of an earthquake can be determined from this line.

The size, or *magnitude*, of an earthquake is determined by the amount of energy it releases. Scientists use the Richter magnitude scale to rate the amount of energy released by an earthquake. The higher the Richter number, the greater the amount of energy that is released.

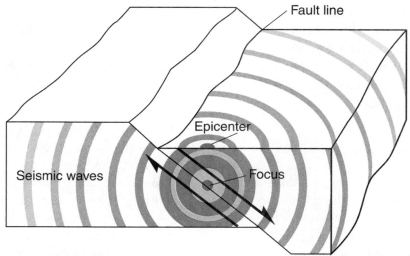

Figure 4.4-2. An earthquake epicenter is located on Earth's surface directly above the focus. The focus is the point underground where the earthquake occurs.

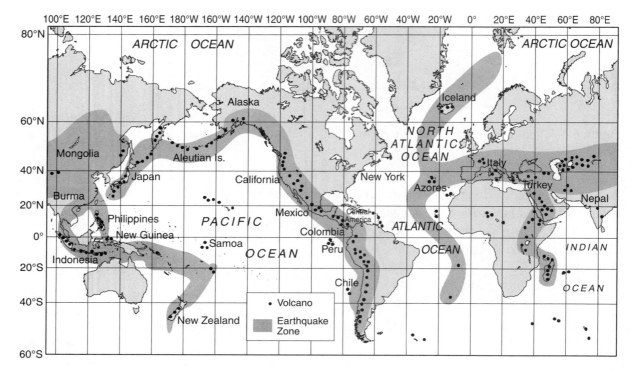

Figure 4.4-3. Can you see a pattern when earthquake zones and volcanoes are plotted on a world map?

There are more than a million earthquakes each year. Earthquakes with a magnitude of 2 or less are not usually felt by people. Each year Earth experiences several thousand earthquakes with magnitudes of 4.5 or greater. Most seismographs around the world can detect these earthquakes. Only about one earthquake each year has a magnitude of 8 or greater.

Earthquake Damage

Contrary to the beliefs of some people, earthquakes do not cause the ground to open up and swallow people. Yes, earthquakes are more likely to occur along plate boundaries, but they can occur anywhere. They are a potential hazard to people throughout most of the United States. A *hazard* is a dangerous event that can cause death, injury, and property damage. Earthquake education and preparedness can reduce the risk of injury and the loss of life.

The following list provides examples of things you can do to prepare for an earthquake:

- Be sure your home meets earthquake standards. Avoid living in or building a home on a steep mountainside or on loose soil.
- Know how to shut off your home's gas, electricity, and water supply.
- Store an emergency box containing food, water, battery-operated radio, flashlight, and medical supplies.

- Keep a list of emergency phone numbers handy.
- Know where to get medical assistance. Be familiar with first-aid procedures. Be prepared to help people with special needs.

Review this list with your family. In addition, discuss what to do when an earthquake occurs. It is always best to stay away from things that can fall. If you are indoors, seek protection under a strong desk or table, or stand in a doorway that will protect you from a falling ceiling. If you are outside, stay clear of trees, electrical poles and wires, and buildings. An earthquake usually lasts only a minute or two, so knowing what to do and rehearsing what to do will better prepare you and keep you safe.

Activities

1. The earthquake that hit San Francisco in 1906 caused devastating damage to the city and disrupted the lives of many people. San Francisco was completely destroyed. Use library resources or the Internet and read about the San Francisco earthquake of 1906. It is worth reading about the San Francisco earthquake to learn how earthquakes cause damage. It also helps to prepare for future earthquakes. Make a list of the dangers that people need to consider when an earthquake strikes.

2. Construct a United States "earthquake watch" display. You will need a map showing the latitude and longitude of the United States, which can be obtained at *http://www.lib.utexas.edu/maps/united_states/ united_states_pol02.jpg*, and current earthquake data, which can be obtained at *http://www.earthquake.usgs.gov/eqcenter/* Obtain a list of earthquakes of magnitude 3.0 or greater that occurred in the United States during the last week. Place a dot on the map to show the location of each earthquake epicenter. After plotting the earthquakes, answer these questions:

a) What state had the most earthquakes?

b) How many earthquakes had a magnitude of 4.0 or greater?

SKILL EXERCISE—*Plotting Earthquakes by Latitude and Longitude*

New York State has a relatively low number of earthquakes. However, New York State has had a few moderate earthquakes in the last 300 years. The table below lists 16 moderate earthquakes that occurred in New York State. Get a copy of a New York State map at *http://www.lib.utexas.edu/maps/states/new_york.gif* Plot the position of each earthquake listed in Table 4.4-1.

Table 4.4-1. Major Earthquakes in New York State

Date	Latitude (° North)	Longitude (° West)	Magnitude
Dec. 18, 1737	40.60	73.80	5.0
Mar. 12, 1853	43.70	75.50	4.8
Dec. 18, 1867	44.05	75.15	4.8
Dec. 11, 1874	41.00	73.90	4.8
Aug. 10, 1884	40.59	73.84	5.3
May 28, 1897	44.50	73.50	NA
Mar. 18, 1928	44.50	74.30	4.5
Aug. 12, 1929	42.84	78.24	5.2
Apr. 20, 1931	43.50	73.80	4.5
Apr. 15, 1934	44.70	73.80	4.5
Sep. 5, 1944	45.00	74.85	6.0
Sep. 9, 1944	45.00	74.85	4.0
Jan. 1, 1966	42.84	79.25	4.6
Jun. 13, 1967	42.84	78.23	4.4
Oct. 7, 1983	43.97	74.25	5.1
Apr. 20, 2002	44.50	73.50	5.1

Questions

1. How many of the earthquakes were in the New York City area?

2. Where did the earthquake with the highest magnitude occur?

Interesting Facts About Tsunamis

On December 26, 2004, a major earthquake struck close to the Indonesian island of Sumatra. The underwater earthquake measured between 9.1 and 9.3 on the Richter scale. It was caused by a 30-m (100-foot) vertical movement of the ocean's floor along a 1000-km (620-mile) section of the Indian-Eurasian plate boundary. The earthquake generated a giant wave called a tsunami. A **tsunami** is a series of large ocean waves generated by an earthquake on the ocean floor. Underwater volcanic eruptions and large underwater landslides can also produce tsunamis.

Hardly noticed by ships at sea, a tsunami travels as several waves across the ocean. In the open sea, the waves are usually less than 30 cm high (1 foot) and travel at 800 km/hr (500 miles/hour). Near shore, the ocean floor slows the wave to 30–100 km/hr (18–62 miles/hour). The faster-moving top of the wave may build to a wave height of 15 meters (50 feet) or more.

Unlike a normal wave along the beach, a tsunami forces water inland, knocking down structures and carrying water great distances in low-lying coastlines. The water then recedes back to the ocean, carrying debris from just about everything it destroyed. Usually, there are several major waves. Within several hours, the Indonesian tsunami slammed the coastline of 11 countries, leaving more than 225,000 people dead or missing. It was the most destructive tsunami in history.

Questions

1. Movement of Earth's crust along plate boundaries produces
   (1) fronts   (3) hurricanes
   (2) tides   (4) earthquakes

2. The diagram below shows a cross section of some rock layers in Earth's crust.

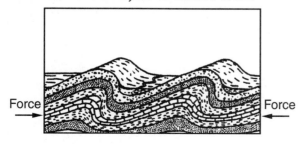

The rock layers are
   (1) faulted   (3) eroded
   (2) folded   (4) expanded

3. A recent earthquake struck California. Cities across the United States recorded the same magnitude for the earthquake. This is because the magnitude of an earthquake depends on
   (1) the distance from the earthquake
   (2) the depth of the earthquake
   (3) when the earthquake occurs
   (4) the amount of energy the earthquake releases

4. The vibrations produced by an earthquake are called
   (1) sound waves
   (2) tidal waves
   (3) seismic intensity
   (4) seismic waves

5. A 3.8-magnitude earthquake occurred 6.5 km underground near Bar Harbor, Maine on October 3, 2006. The point on Earth's surface directly above where this earthquake occurred is called the earthquake's
   (1) focus
   (2) seismic point
   (3) epicenter
   (4) volcano

Thinking and Analyzing

1. Use the table below and your knowledge of science to answer the following questions. The table shows information about some major world earthquakes.

Some Major World Earthquakes

Year	Location	Deaths	Magnitude
1556	Shaanxi, China	830,000	NA
1886	Charleston, South Carolina	60	NA
1906	San Francisco, California	700	8.3
1920	Gansu, China	200,000	8.6
1923	Tokyo, Japan	143,000	8.3
1960	Southern Chile	5,000	9.5
1964	Alaska	131	9.2
1976	Tangshan, China	255,000	8.0

NA: Information is not available.

(a) Which earthquake released the greatest amount of energy?
(b) How can scientists in the United States determine the date, time, and size of an earthquake that occurs in the middle of China?

(c) In 1886, a major earthquake occurred in Charleston, South Carolina. Why was this earthquake different from most other major earthquakes?

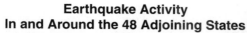

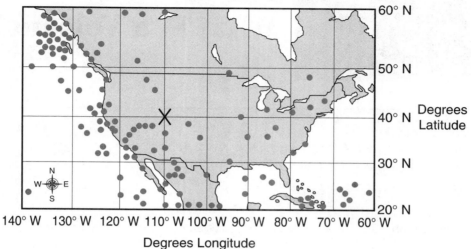

Degrees Longitude

2. Use the map above and your knowledge of science to answer the following questions. The map shows earthquake activity in and around the adjoining 48 states. Earthquake activity is indicated by dots.

(a) State one reason why there are more earthquakes in the western section of the area shown on the map.

(b) According to the map, what is the latitude and longitude of the location at letter X? (Your answer must include a value, unit, and direction for each.)

(c) List two actions that residents of the West Coast might include in an earthquake emergency preparedness plan.

4.5 What Is a Volcano?

Objectives

Describe the types of material released from a volcano.

Describe how a volcano forms.

Explain how volcanoes are distributed on Earth.

Terms

volcanic (vahl-KAN-ic) **activity:** processes related to the movement of liquid rock underground (intrusive) and on Earth's surface (extrusive)

volcano (vahl-KAY-noh): an opening in Earth's surface where liquid rock is released

ash: fine dust particles produced by a volcanic eruption

cinders: coarse rock particles produced by a volcanic eruption

volcanic cone: a volcanic mountain

crater: a depression at the top of a volcanic cone

caldera (kal-DER-uh): a basin formed by the collapse of a volcanic mountain

Volcanic Activity

Volcanic activity describes the processes related to the movement of liquid rock underground (intrusive) and on Earth's surface (extrusive). Underground heat and pressure can melt rock material in the lower crust, forming a liquid rock called *magma*. (See Figure 4.5-1.) Pressure and density differences between magma and the surrounding rock cause magma to rise toward Earth's surface. If an opening or **volcano** exists, magma may be released. At the surface, the liquid rock is called *lava*. White-hot lava has a temperature greater than 1000°C. *Stop and Think:* Do you

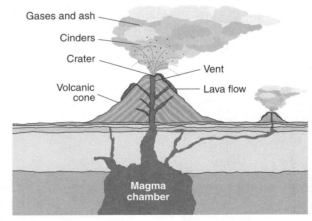

Figure 4.5-1. A cross section of a typical volcano showing some of its features.

remember what type of rock forms when magma and lava solidify? *Answer:* Magma and lava form igneous rock.

Volcanoes also release large quantities of gases. The most common gases released are water, carbon dioxide, and sulfur dioxide. Other gases released from volcanoes are hydrogen, hydrogen sulfide, and hydrochloric acid.

Explosive volcanic eruptions eject lava high into the air. (See Figure 4.5-2.) The hot lava cools quickly to form solid particles. Fine, dust-size particles are called **ash**. Winds in the upper atmosphere can carry ash around Earth. Large amounts of ash particles in the atmosphere can block the sun's radiation and cause a decrease in Earth's temperature. Cinders are coarser rock material that fall out of the air quickly around the volcano.

A **volcanic cone** is a mountain produced by the accumulation of lava and cinders around its opening. Cinders fall from the air and build a steep-sloped cone. Lava hardens around the opening and builds a moderately sloped cone. Underground channels called *vents* carry the magma to a depression, or **crater**, usually at the top of a volcanic cone. Most volcanic activity in the United States occurs in Hawaii and Alaska. Occasionally, the volcanic mountains in Washington, Oregon, and California become active.

Location of Volcanoes

Volcanoes tend to be located near plate boundaries and in earthquake zones. A narrow belt of volcanoes around the Pacific Ocean is called the "Ring of Fire" because about 60 percent of Earth's more than 600

Figure 4.5-2. The volcanic eruption of Mount St. Helens on May 18, 1980, ejected a cloud of ash, cinders, and gases high into the atmosphere.

active volcanoes are found there. (Refer back to Figure 4.4-3 on page 104.) Most of these volcanoes are associated with trenches and subduction zones on the boundary of the Pacific Plate. As an oceanic plate slides down and under a continental plate, friction and pressure produce heat that melts the rock and creates magma. The magma rises and often produces a volcano. The Aleutian Islands in Alaska and the Andes mountains in South America are examples of volcanoes near subduction zones. A large number of volcanoes are also found along plate boundaries in southern Europe and Africa.

Many volcanoes also occur along the mid-ocean ridge. (Look back at Figure 4.4-3 on page 104.) Underwater volcanic eruptions produce the mid-ocean ridge. If the volcanoes grow high enough, a volcanic island is produced. Iceland and the Azores in the Atlantic Ocean are volcanic islands located along the Mid-Atlantic Ridge.

Fewer volcanoes form at *hot spots* in the middle of crustal plates. Think of a hot spot as a hole in the middle of a plate. The hole allows magma to reach the surface and form a small group of volcanic mountains. The Hawaiian Islands are produced by a hot spot under the middle of the Pacific Plate.

Stop and Think: Why are there no active volcanoes in Florida? *Answer:* Florida does not have any active volcanoes because it is not near the edge of a crustal plate, and there are no hot spots or magma forming beneath the area.

Examples of Volcanic Eruptions in the United States

Mount St. Helens, a volcanic mountain in Washington State's Cascade Mountain Range, is about 160 km (100 miles) from the Pacific Ocean. On May 18, 1980, in a matter of a few minutes, an explosive eruption leveled everything over an area of 150 square miles. It is estimated that 800,000 tons of ash fell on Yakima, Washington, 135 km (85 miles) east of the volcano. Upper air winds carried volcanic dust around Earth for more than two years. The 1980 eruption of Mount St. Helens was its first since 1854.

Crater Lake is located in southern Oregon along the Pacific Coastal Mountain Range. It was created when Mount Mazama, 3660 m (12,000 feet) high, collapsed during a huge eruption 7700 years ago. The collapsed volcano produced a basin, or **caldera**, which later filled with rain and the water from melted snow, producing a lake. Crater Lake is 597 m (1958 feet) deep and is the deepest lake in the United States. More recent eruptions have produced Wizard Island, a small cinder cone in the lake.

Kilauea is an active volcano in the Hawaiian Islands. It is considered a "quiet volcano" because it does not have explosive eruptions. It produces large quantities of lava and tends to build a more gently sloped volcanic cone, called a shield cone. Since 1983, Kilauea has been erupting continuously, making it one of Earth's most active volcanoes.

Activity

Use library resources or the Internet and research any one of the following unique volcanic eruptions. Write a brief description of the eruption. Be sure to include the location of the volcano and at least three major items about the eruption.

Krakatoa, 1883
Pelee, 1902
Paricutin, 1943
Vesuvius, 79 AD

SKILL EXERCISE—*Plotting Volcanoes by Latitude and Longitude*

Volcanoes are not randomly scattered on Earth. Most volcanoes are found near crustal plate boundaries. Therefore, they are grouped into zones. The five volcanic zones on Earth are:

1. The Ring of Fire around the Pacific Ocean, where plates are converging

2. Southern Europe, where plates are converging

3. Africa, where plates are diverging

4. The mid-ocean ridge, where plates are diverging

5. Hot spots in the middle of plates

Locate the five volcanic zones on the map. (See Figure 4.4-3 on page 104.)

Questions

1. The table below gives the latitude and longitude of five volcanoes.

Latitude and Longitude of 5 Volcanoes		
Volcano Number	**Latitude**	**Longitude**
1.	19° N	98° W
2.	6° S	105° E
3.	38° N	15° E
4.	64° N	18° W
5.	19° N	150° W

Use latitude and longitude and the map to find the location of each volcano listed in the table above. Construct a chart listing the general location and volcano zone of each volcano.

2. Obtain the latitude and longitude of five recent volcanic eruptions from the Smithsonian Institution/USGS Weekly Volcanic Activity Report at *http://www.volcano.si.edu/reports/usgs/* In which volcano zone is each eruption located?

Interesting Facts About the Eruption of Tambora in 1815

Tambora, a 4000-m (13,000-foot) high volcano in Indonesia, had a violent eruption in April 1815. The 5-day eruption blew about 100 cubic miles of dust, ash, and cinders into the atmosphere.

Winds carried suspended volcanic dust completely around Earth for at least 2 years. The dust produced colorful sunrises and sunsets, and decreased the amount of solar radiation reaching Earth. Many scientists believe the floating dust caused Earth's temperature to drop a few degrees, causing much distress to farmers in New England. In the summer of 1816, frost and snow destroyed crops several times in New England, causing a severe food shortage. Europe suffered severe cold spells and famine. The year 1816 is known as "the year without a summer." It is believed the eruption of Tambora was the cause of this extraordinary weather.

Questions

1. What type of rock does volcanic activity produce?
   (1) sedimentary rock
   (2) metamorphic rock
   (3) igneous rock
   (4) a mixture of sedimentary, igneous, and metamorphic rock

2. The bowl-like depression at the top of a volcano is called a
   (1) vent   (3) cone
   (2) crater   (4) caldera

3. Steep-sloped volcanic cones are formed by
   (1) volcanic gases (3) volcanic cinders
   (2) volcanic ash   (4) volcanic lava

4. Magma rises to Earth's surface because
   (1) it is less dense than the surrounding rock
   (2) it is more dense than the surrounding rock
   (3) it has the same density as the surrounding rock
   (4) of a decrease in air pressure

5. Most volcanoes around the Pacific Ocean are associated with
   (1) subduction zones along plate boundaries
   (2) mid-ocean ridges
   (3) hot spots
   (4) meteor impact zones

Thinking and Analyzing

1. Base your answers to the following questions on the diagram at the right and on your knowledge of science. The diagram shows a tectonic plate boundary.
(a) The volcano in the diagram is associated with what type of plate boundary?
(b) What ocean floor feature is associated with this type of plate boundary?
(c) Give an example of where this type of plate boundary can be found.

2. Base your answers to the following questions on the map below and on your knowledge of science. The map shows the relationship between volcanoes and plate boundaries.
(a) Volcano A is associated with what ocean floor feature?
(b) Volcano B is associated with what type of plate boundary?

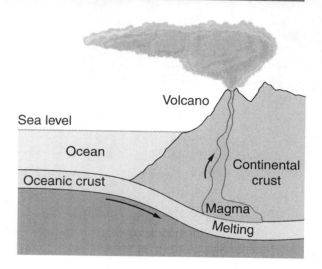

(c) What is the "Ring of Fire"? (Hint: It has nothing to do with the song by Johnny Cash.)

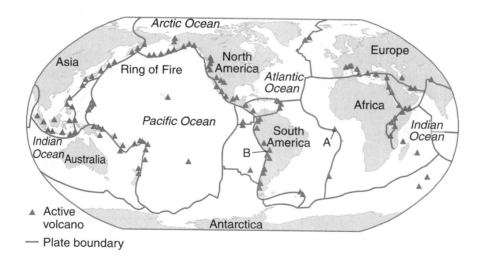

How Do Mountains Form?

Objectives
Describe the different ways mountains form.

Describe how mountains can change over long periods of time.

Terms
landform: a large land feature defined on the basis of how high it is, the steepness of its slope, and the type of rock it is made of; examples of common landforms are mountains, plains, and plateaus

mountain: a landform that is significantly higher than the land surrounding it

plain: a large, generally flat landform at a low elevation

plateau (pla-TOE): a large, generally flat landform at a high elevation

---

Landforms

A **landform** is a large land feature defined on the basis of how high it is, the steepness of its slope, and the type of rock it is made of. The three major landforms are mountains, plains, and plateaus.

A **mountain** is a landform that is significantly higher than the land surrounding it. (See Figure 4.6-1.) Mountains have steep slopes and are usually composed of igneous or metamorphic rock. Each individual mountain, like Pike's Peak in Colorado, has a peak or *summit* at the top.

Mountains cover about one-fifth of Earth's surface. Most mountains are part of a group of mountains, called a *mountain range*. You are probably familiar with large mountain ranges such as the Rocky Mountains and the Appalachian Mountains. However there are many smaller mountain ranges such as the Big Horn Mountains in Wyoming and the Catskill Mountains in New York State. There are also single mountains within a mountain range, such as Mount St. Helens in Washington and Mount McKinley (Denali) in Alaska. Each mountain range is different in formation and structure from every other mountain range.

A **plain** is a large, flat landform at a low elevation. The land is composed of flat layers of sand or sedimentary rock close to sea level. The eastern coastal region of the United States, from New York to Florida, is part of a large coastal plain, called the Atlantic Coastal Plain.

Figure 4.6-1. The Grand Tetons in Wyoming project high above the surrounding land.

A **plateau** is a large, flat landform at a high elevation. It usually has a horizontal rock structure. A plateau is level, but it may become hilly if streams have cut into it. The Allegheny Plateau along the Pennsylvania-New York border is an example of a plateau cut by streams.

Mountain-Building Processes

Tectonic forces, volcanoes, and erosion are all processes that form mountains. Most mountain-building processes are associated with the movement of tectonic plates. Volcanic mountains can be produced by magma pushing the land up, or by the accumulation of cinders or lava on Earth's surface. Erosion forces, such as running water, glaciers, and groundwater can cut into a plateau and produce mountains. There are several different types of mountains, based on how they were formed.

Folded Mountains

Sideways (lateral) pressure causes the land to bend up and down in a wave-like appearance. To help you understand this process, slide a small loose rug into a wall with your foot. The up-and-down folding this causes in the rug is a model of how mountain folds are produced on land. Folded mountains are usually produced from sedimentary rock, and consist of elongated (stretched-out, slender) mountains and parallel valleys. The Appalachian Mountains in eastern Pennsylvania are folded mountains. (See Figure 4.6-2A on page 118.)

Fault-Block Mountains

Fault-block mountains are produced by the upward thrust of a large section of land along a fault line. These mountains have a

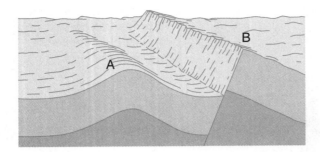

Figure 4.6-2. (A) The wave-like appearance of the rock is associated with folded mountains. (B) The "tilted" structure with a steep slope on one side and a more gradual, gentle slope on the other side is associated with fault-block mountains.

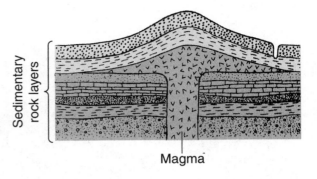

Figure 4.6-3. Dome mountains are created when land is pushed up by trapped volcanic magma.

steep slope on one side, and a more gradual slope on the opposite side. The Sierra Nevada Mountains in California, and the Great Basin and Range of Utah, are fault-block mountains. (See Figure 4.6-2B.)

Dome Mountains

Large reservoirs of magma underground can force the overlying land to lift up into an arch or dome. (See Figure 4.6-3.) The surface bulges up, producing a generally circular mountain range. The dome rarely exceeds 240 km (150 miles) in diameter. The Adirondack Mountains in New York State and the Black Hills in South Dakota are dome mountains.

Volcanic Mountains

Lava, cinders, or some combination of both produce volcanic mountains. The type of material a volcano releases determines the shape of the mountain produced. Some volcanic eruptions produce lots of lava;

other, more violent, eruptions produce lots of cinders. Liquid lava spreads out from a volcano, producing a broad-based volcanic mountain with a relatively gentle slope. Falling cinders from explosive eruptions produce a narrow-based mountain with a relatively steep slope. (See Figure 4.6-4.) The Hawaiian Islands are volcanic mountains produced primarily by lava. ***Stop and Think:*** Describe the mountains that make up the Hawaiian Islands. ***Answer:*** The Hawaiian Islands are volcanic mountains with a broad base and a relatively gentle slope.

Erosion Mountains

Sometimes tectonic forces lift the land and produce an elevated flat surface, or plateau. Rivers and streams erode the plateau, forming valleys and leaving higher elevations in the form of mountains between the valleys. The mountains formed by erosion in a mountain range usually have flat tops or ridges at about the same height. The Catskill Mountains in New York State were formed by erosion.

Figure 4.6-4. Sunset Crater is a volcanic mountain in Arizona. Volcanic eruptions produced this cinder cone mountain almost 1000 years ago.

Mountain Stages

Mountain-building forces are the direct opposite of erosion forces. Plate tectonics and volcanoes lift the land and cause mountains to build up. Weathering and erosion wear mountains down to flat plains at sea level. If mountain-building forces are greater than erosion forces, the mountain's height increases. However, if erosion forces are greater than mountain-building forces, the mountain's height decreases. The Andes Mountains in South America are growing, and the Appalachian Mountains in Pennsylvania are wearing down. ***Stop and Think:*** Describe the relationship of mountain-building and erosion forces in the Appalachian Mountains. ***Answer:*** In the Appalachian Mountains, erosion forces are greater than mountain-building forces.

Given enough time and the lack of mountain-building forces, mountains eventually wear down to ground level. As they change, they appear to go through stages. (See Figure 4.6-5.) The stages are

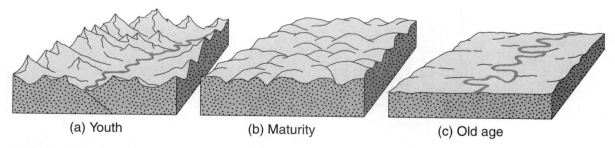

(a) Youth   (b) Maturity   (c) Old age

Figure 4.6-5. Mountains in three stages of development: (a) youth, (b) maturity, and (c) old age.

based on mountain characteristics, not on age in years. As a 50-year-old person can look younger than a 40-year-old person, so can an older mountain range be at a younger stage than a mountain that is actually years younger.

Youth Stage

Mountains in the youth stage have the following characteristics: high elevations, earthquakes or active volcanoes, sharp rocky peaks, steep slopes, landslides, and narrow river valleys. Examples of mountains in the youth stage are the Himalayas, Alps, and Rocky Mountains.

Mature Stage

Mature mountains have the following characteristics: moderate elevations, few earthquakes, no active volcanoes, rounded peaks, U-shaped river valleys, and few landslides. The Appalachian Mountains and the Adirondack Mountains are mature mountains.

Old-Age Stage

Mountains in the old-age stage are worn down close to ground level. Their land surface is almost flat with some rolling hills. River valleys are almost nonexistent and earthquakes are rare. The area north of New York City is an example of an old-age stage mountain.

Interesting Facts About the Himalayas

About 80 million years ago, the Indian Plate was about 6400 km (3975 miles) south of the Eurasia Plate. The Indian Plate was slowly moving north. Its motion eventually caused the two landmasses to collide about 50 millions years ago. Along the colliding boundary, the Himalayas, Earth's highest mountain range, formed. Many of the world's tallest mountains are located in the Himalayas.

Mountain	Height	
	(meters)	(feet)
Everest	8850	29,035
K2	8611	28,250
Kanchenjunga	8598	28,208
Lhotse	8511	27,923
Makalu I	8481	27,824

The Indian Plate is still moving at 67 mm/yr (2.6 in./yr), causing the Himalayas to rise about 5 mm/yr (0.2 in./yr). If this rate of motion continues, the Himalayas will rise another 0.5 meter (1.6 feet) in the next 100 years.

Activities

1. Perform each activity below and determine which mountain-building process it demonstrates. In your notebook, draw a diagram of each model, write a description of each, and state the type of mountain-building process each demonstrates.

(a) Place a loose rug on a hard floor. Slowly push the rug into the base of the wall with your foot.

(b) Take a handful of dry sand and slowly let it fall on a flat surface. Repeat this several times to build up a pile of sand.

(c) Place a balloon in a shallow pan and cover it with 5 cm of sand or soil. Leave the lip of the balloon uncovered and hanging to the side of the pan. Place the pan near the edge of a table so you can slowly blow air into the balloon.

2. Make a copy of a map showing the outline of the United States or print one from *http://www.eduplace.com/ss/maps/*. Locate and label each of the following mountain regions on your map.

(a) Adirondack Mountains

(b) Appalachian Mountains

(c) Black Hills

(d) Cascade Range

(e) Catskill Mountains

(f) Ozark Plateaus

(g) Rocky Mountains

(h) Sierra Nevada Mountains

Questions

1. Which is a major landform?
   (1) an ocean   (3) a mountain
   (2) a stream   (4) a city

2. Which process tends to destroy mountains?
   (1) intrusive volcanic activity
   (2) extrusive volcanic activity
   (3) erosion
   (4) earthquakes

3. Mature mountains are best described as having
   (1) high elevations, steep slopes, and many earthquakes
   (2) moderate elevations, landslides, and narrow river valleys
   (3) moderate elevations, moderate slopes, and few earthquakes
   (4) low elevations, rolling hills, and no earthquakes

4. The Alps are mountains in the youth stage. Which statement correctly describes the forces acting on the Alps?
(1) Mountain-building forces and erosion forces are equal.
(2) Mountain-building forces have stopped and erosion forces have started.
(3) Mountain-building forces are greater than erosion forces.
(4) Erosion forces are greater than mountain-building forces.

5. The three major landforms are
(1) mountains, plains, and plateaus
(2) mountains, oceans, and rivers
(3) mountains, hills, and valleys
(4) highlands, lowlands, and midlands

Thinking and Analyzing

1. Base your answers to the questions on the map below and on your knowledge of science. The map shows three landscape regions in the Mid-Atlantic region.

(a) Describe the landscape of the Mid-Atlantic Coastal Plain.
(b) What evidence on the map indicates that the Adirondack Mountains are dome mountains?
(c) How does a plateau differ from a plain?

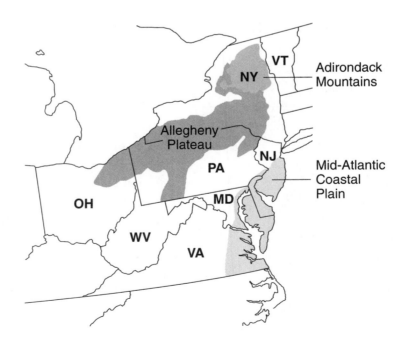

2. Base your answers to the questions on the reading passage and the diagram below, and on your knowledge of science.

Sierra Nevada Mountains

The Sierra Nevada Mountains are located in California. The mountain range stretches 650 km (400 miles) along California's eastern border. West of the Sierra Nevada Mountains is California's Central Valley, and on the east side is the Basin and Range Province, which consists of mountains and valleys. If you travel from west to east, the elevation gradually increases until you reach the highest elevation. The elevation decreases rapidly on the east side of the mountains. Many earthquakes are associated with the continued uplift of the Sierra Nevada Mountains.

(a) On which side of the Sierra Nevada Mountains do you expect most earthquakes to occur? Why?
(b) What type of mountains are the Sierra Nevada Mountains?
(c) Describe two features (not mentioned in the passage above) of mountains in their youth stage of development.

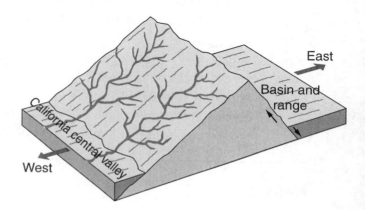

What Is a Topographic Map?

Objectives

Identify the type of information found on a topographic map.

Identify different types of land features using a topographic map.

Determine elevations and distances on a topographic map.

Terms

topographic (tah-puh-GRAF-ihk) **map:** a map that shows the form and shape of the physical land features

contour (KAHN-toor) **lines:** lines connecting points of equal elevation

contour interval: elevation difference between two adjacent contour lines

profile: a side view, or cross section, of a cutaway section of Earth's surface

---

Maps

A *map* is a flat model showing a portion of Earth's surface. Different types of maps show specific information about Earth's surface and provide useful information. Weather maps show weather conditions and provide important information to airplane pilots, boaters, and skiers. Road maps show the location of roads, towns, and cities, and assist in everyday travel for drivers.

Topographic maps show the form and shape of physical surface features. The physical surface features include mountains, valleys, hills, plains, streams, and lakes. Topographic maps are used by engineers when developing roads, by hikers planning a hike, and by anyone interested in knowing the shape of the land.

Features of Topographic Maps

Contour lines are used on topographic maps to show the shape of the land. (See Figure 4.7-1.) Contour lines connect points of equal elevation. They represent land elevations above mean sea level. Mean sea level is the average sea level between low and high tide, and it represents 0 m elevation. The elevation difference between two adjacent contour lines is called the **contour interval**. Often, to make it easier to read the elevation, every fifth contour line is darker, and its elevation is labeled. This is called an *index contour* line.

The size of a topographic map is determined by the amount of latitude and longitude shown on the map. *Stop and Think:* On a globe of Earth, which lines are

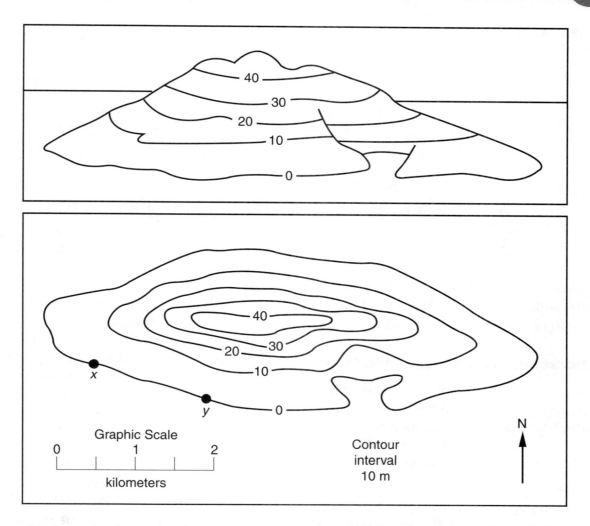

Figure 4.7-1. A simple topographic map of an island. The contour interval is 10 meters and the height of the island is about 45 meters. The distance from point *x* to point *y* is about 1.5 kilometers.

latitude lines and which are longitude lines? **Answer:** Latitude lines run east-west, parallel to the equator. Latitude lines measure distance from the equator. Longitude lines run north-south, from the North Pole to the South Pole. Longitude lines measure distance from the Prime Meridian.

In addition to showing the shape of the land, topographic maps use colors and symbols to show other features. Blue is used to indicate bodies of water, such as lakes and rivers. Features such as roads, buildings, railroad tracks, and cemeteries are shown in black. Green is used to show areas of dense vegetation, such as woodlands. Contour lines are brown. Different types of symbols are used on topographic maps to show many natural

Figure 4.7-2. Some symbols used on topographic maps.

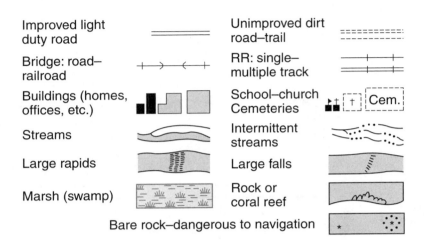

Improved light duty road	Unimproved dirt road–trail
Bridge: road–railroad	RR: single–multiple track
Buildings (homes, offices, etc.)	School–church Cemeteries
Streams	Intermittent streams
Large rapids	Large falls
Marsh (swamp)	Rock or coral reef

Bare rock–dangerous to navigation

and artificial features. Figure 4.7-2 shows some of these symbols.

Information on Topographic Maps

The United States Geological Survey (USGS) prints topographic maps of areas in the United States. Much information about a topographic map can be found in its margins. Some of this information includes:

- *Map name*: The name of a topographic map is taken from a major geographic or artificial feature found on the map. A typical name may come from a river, mountain, city, landform, or some other large feature. For example, the "Mt. Shasta" California map is named for Mt. Shasta, the largest physical land feature on the map.
- *Year of production or revision*: The year of production or revision indicates the year the map was first published or last revised.
- *General location in state*: A small black box within a state outline

indicates the geographic location of the map within a state.

- *Map scale ratio*: The map scale ratio is written in numerical form; for example, 1:62,500 or 1:24,000. This represents the ratio of map units to Earth's surface units. The ratio of 1:24,000 means 1 cm on the map equals 24,000 cm (240 meters) on Earth, or 1 in. on the map equals 24,000 in. (2000 feet) on Earth.
- *Line graphic scale*: The graphic scale is a line divided into segments and labeled with the distance it represents. (See the graphic scale on Figure 4.7-1.)
- *Contour Interval*: Typical contour intervals for a topographic map are 5, 10, or 20 meters, or 10, 20, or 50 feet.
- *Difference between magnetic north and true north*: The difference between magnetic north (MN) and true north (GN—geographic north) is indicated in a diagram showing the degree difference between the two directions. Magnetic north is the

direction indicated by a compass needle, and true north is the direction of the North Pole.

- Latitude and Longitude Labels: The latitude and longitude of a topographic map are labeled in each of the map corners, and along each of the four sides.

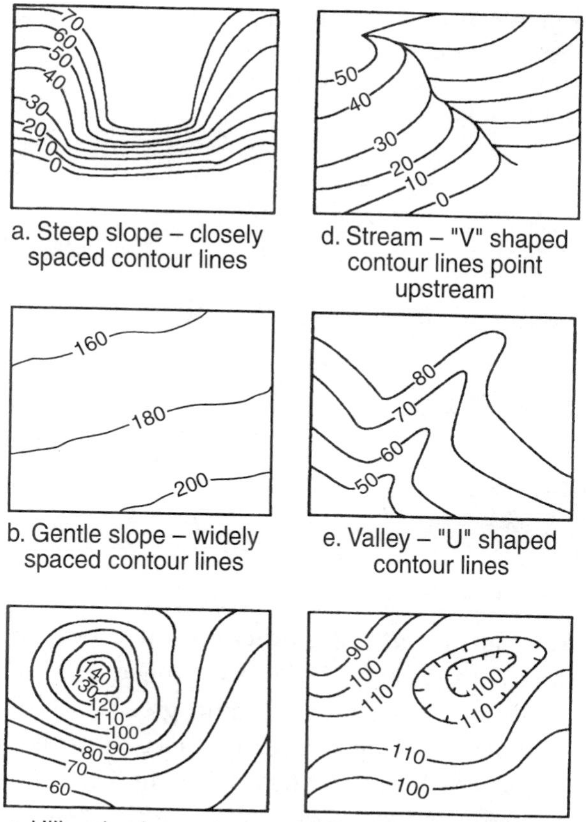

Landscapes

Landscapes are shown by the shape and spacing of contour lines. (See Figure 4.7-3 a–f.) For example:

- Closely spaced contour lines indicate a steep slope (a).
- Widely spaced contour lines indicate a gentle slope or nearly flat area (b).
- Circular contour lines indicate a hill or mountain (c).
- Contour lines that cross a stream or river make a V shape that points upstream (d).
- U-shaped bends in contour lines indicate a wide, deeply eroded valley (e).
- Hachure marks (short inward pointing lines) on a contour line indicate a depression (f).

a. Steep slope – closely spaced contour lines

b. Gentle slope – widely spaced contour lines

c. Hill – circular contour lines

d. Stream – "V" shaped contour lines point upstream

e. Valley – "U" shaped contour lines

f. Depression – hachure marks on contour lines

Figure 4.7-3. You can learn to recognize various landscapes on a topographic map from the shape of the contour lines.

Profiles

A **profile** is a side view or cross section of a cutaway section of Earth's surface. A profile shows how the land goes up and down as if you were walking a straight line on Earth's surface. On a topographic map, a profile of the landscape can be projected from a straight line on the map. (See Figure 4.7-4.)

By drawing a profile between two points on a topographic map, you get a true picture of the shape of the land you would cross if you traveled from one point to another point. For example, in Figure 4.7-4, if you were to walk from point M to point N in a straight line, you would have to climb over a hill.

Figure 4.7-4. Projecting a profile from a topographic map gives you a side view, or cross section, of the landscape.

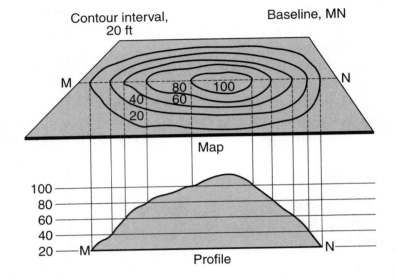

Activities

1. Make a cardboard model of the topographic island in Figure 4.7-1 on page 125. You will need a stiff piece of cardboard about 20 cm × 20 cm, glue, green and blue crayons, and a black marking pen. Cut a piece of cardboard 10 × 20 cm. This piece will be your base. Cut out 5 additional pieces in the shape of each contour line on the map. Glue the largest contour piece in the center of the base. Continue gluing all the pieces on top of each other in the same layout as the map. Color the water blue and the island green. Label each contour line in your model.

2. Produce a contour map of a rock. You will need a 4- to 8-liter pail, a rock about 15 cm in width, a ruler, dyed water, plastic wrap (or sheet of clear plastic), and a marking pen. The pen must be able to mark the plastic wrap or sheet. Place the rock in the pail so the flattest side is down and add 1 cm of water. Stretch the plastic wrap over the top of the pail and, looking straight down at the rock, draw an outline on the plastic wrap where the rock and water touch. Carefully pull back just enough plastic wrap so you can add another 1 cm of water. Replace the plastic wrap and again draw an outline where the rock and water touch. Repeat this procedure until the rock is covered by water. Complete your contour map by labeling the contour lines 0, 1, 2, 3, etc. Cut a 20-cm square around your plastic wrap map and tape it to a page in your notebook.

3. Use the Internet to obtain a portion of a USGS topographic map. Go to Topozone at *http://www.topozone.com/ viewmaps.asp.* Select North Carolina, select Watauga County, and select Boone Golf Course. Study the map carefully. You should be able to find mountains, a valley, flat lands, and steep slopes. What other features are visible? Using the Topozone Web site, see if you can find other land areas near your home.

SKILL EXERCISE—*Finding True North*

A compass consists of a magnetized metal needle that freely pivots above a circular dial that is labeled with the major geographic direction points. The most commonly labeled points are north, northeast, east, southeast, south, southwest, west, and northwest. (See Figure 4.7-5a below.)

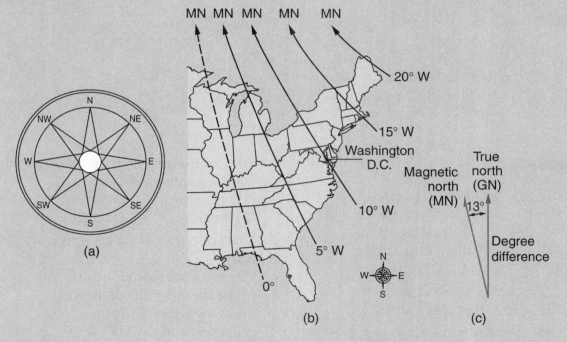

Figure 4.7-5.
Earth's iron-rich core acts much like a giant bar magnet. At one end of the giant magnet is magnetic north. When allowed to swing freely and rest, the compass needle points to magnetic north (MN). True north (usually labeled GN for geographic north) is located at the North Pole.

Magnetic north and true north are at different locations; therefore, they are in different directions. In Virginia, magnetic north is between 3 and 10 degrees *west* of true north. (See Figure 4.7-5b.) When accuracy of direction is required, this difference must be taken into account.

In Virginia, you can determine true north using a compass and a topographic map. First, from the margin of a local topographic map, obtain the difference between true north and magnetic north. (See Figure 4.7-5c.) Now, use your compass to locate magnetic north. Move eastward from magnetic north the number of degrees indicated in the margin on the topographic map. This is the direction of true north.

Questions

1. Explain where in the United States a compass needle points to true north.

2. What is the degree difference between magnetic north and true north in Washington, D.C.?

3. In Virginia, magnetic north is always
   (1) west of true north
   (2) east of true north
   (3) south of true north
   (4) in the same direction as true north

Questions

1. Circular contour lines indicate a
   (1) hill   (3) depression
   (2) valley   (4) ridge

2. Closely spaced contour lines on a topographic map represent
   (1) flat land   (3) a gentle slope
   (2) a steep slope (4) a lake

3. Distances are determined on topographic maps using
   (1) contour lines
   (2) latitude and longitude
   (3) the scale ratio or graphic scale
   (4) the contour interval

4. Contour lines are used on a topographic map to determine a difference in
   (1) elevation   (3) distance
   (2) latitude   (4) longitude

5. True north on a USGS topographic map is
   (1) on the left side of the map
   (2) on the right side of the map
   (3) at the top of the map
   (4) at the bottom of the map

Thinking and Analyzing

1. Base your answers to the questions on the topographic map below.

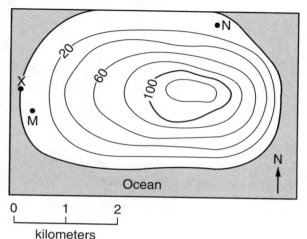

(a) What is the elevation of point X?
(b) What is the distance between points M and N?
(c) The number 100 on the index contour line indicates 100 meters above sea level. What is the contour interval of the topographic map?

2. In your notebook, design a topographic map of an island.
(a) Use 10-meter contour intervals.
(b) The hill on the island has an elevation of 75 meters. Place an X at the top of the hill.
(c) On your map, place a graphic scale that shows 1 cm equals 1 km.

Review Questions

Term Identification

Each question below shows two terms from Chapter 4. One of the terms is defined.
(1) Choose the term that matches the definition.
(2) Describe how the two terms are different. Following each term is the section (in parenthesis) where the description or definition of that term is found.

1. *Trench (4.2) — Mid-ocean ridge (4.2)*
   New ocean floor is created along this plate boundary

2. *Continental crust (4.3) — Ocean floor crust (4.3)*
   Crust composed primarily of granite-like rock

3. *Convergent boundary (4.3) — Divergent boundary (4.3)*
   Type of boundary produced when plates move apart

4. *Epicenter (4.4) — Focus (4.4)*
   The underground location of an earthquake

5. *Folding (4.4) — Faulting (4.4)*
   Bending of rocks

6. *Seismic waves (4.4) — Tsunami (4.4)*
   A series of large ocean waves generated by an earthquake on the ocean floor

7. *Lava (4.5) — Magma (4.5)*
   Melted rock that is underground

8. *Dome mountain (4.6)—Volcanic mountain (4.6)*
   Type of mountain produced by magma that lifts up the rocks above it

Multiple Choice (Part 1)

Choose the response that best completes the sentence or answers the question.

1. Major mountain ranges are formed when crustal plates
   (1) push into each other
   (2) slide past each other
   (3) move away from each other
   (4) break into smaller plates

2. Dust and ash entering the atmosphere as a result of volcanic eruptions can affect Earth's
   (1) tidal activity
   (2) orbital shape
   (3) weather and climate
   (4) rotation and revolution

3. Movement of Earth's crust along plate boundaries produces
   (1) fronts   (3) hurricanes
   (2) tides   (4) earthquakes

4. The map below shows some geologic features located near the west coast of the United States.

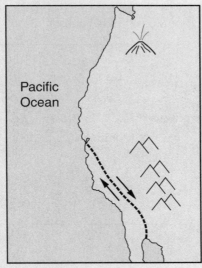

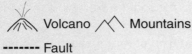

/ⁱ\ Volcano /\/\ Mountains

------- Fault

The arrows on either side of the fault represent
(1) volcanic eruptions
(2) rock formations
(3) the relative movement of air masses
(4) the relative movement of tectonic plates

5. Alaska is more likely than Florida to have an earthquake because
(1) Alaska is younger than Florida
(2) Alaska is in the middle of a crustal plate
(3) Alaska is on the boundary of a crustal plate
(4) Florida has more people

6. If crustal block A, to the left of the fault in the diagram below, suddenly shifted downward several feet, what would most likely occur at location C?

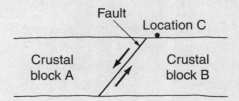

(1) An earthquake would occur.
(2) A volcanic eruption would occur.
(3) A mountain would form.
(4) An ocean would form.

7. Convection currents within Earth's mantle are making the Atlantic Ocean
(1) less salty   (3) wider
(2) cooler   (4) narrower

8. The table below shows the depth of the ocean at various distances from a continent.

Distance from Continent	Ocean Depth
50 km	400 m
100 km	9000 m
150 km	1250 m
200 km	1100 m
250 km	200 m
300 km	950 m

At what distance from the continent is a trench located?
(1) 100 kilometers
(2) 200 kilometers
(3) 300 kilometers
(4) 500 kilometers

9. Many scientists believe that crustal plate movement occurs because of convection cells in Earth's
(1) crust   (3) core
(2) mantle   (4) atmosphere

10. Circular contour lines on a topographic map represent a
(1) hill   (3) depression
(2) valley   (4) ridge

11. On a topographic map, the elevation difference between two adjacent contour lines is called the
(1) contour interval (3) graphic scale
(2) index contour   (4) ratio scale

Thinking and Analyzing (Part 2)

1. Base your answers to the questions on the profile below and on your knowledge of science. The map shows a profile view of the floor of the Atlantic Ocean between South America and Africa.
(a) A rock sample was obtained at position C by drilling into the ocean floor. What type of rock was it?
(b) Radioactive dating was used to obtain the age of the rocks at positions A, B, and C. Which rock sample is the oldest?
(c) What ocean floor feature is found at position D?

2. Base your answers to the questions on the map below and on your knowledge of science. The map shows the movement of volcanic ash after the May 18, 1980 eruption of Mount St. Helens.

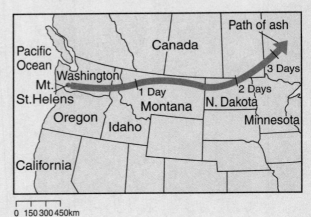

0 150 300 450km

(a) What caused the volcanic ash to move to the east?
(b) What type of rock makes up Mount St. Helens?
(c) State one way the volcanic eruption might affect northern North Dakota.

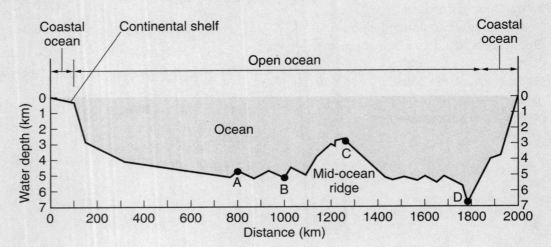

3. Base your answers to the questions on the topographic map below.

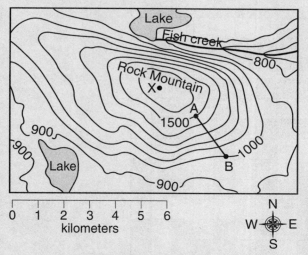

0 1 2 3 4 5 6
kilometers

N
W ⊕ E
S

(a) In what geographic direction is Fish Creek flowing?

(b) Describe the land if you walked from point A to point B.

(c) What is the distance between point A and point B?

4. Base your answers to the questions on the diagram below and on your knowledge of science. The diagram shows a boundary between two tectonic plates.

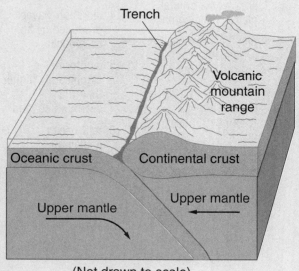

(Not drawn to scale)

(a) What causes the plates to move?

(b) Describe two differences between the oceanic crust and the continental crust.

(c) What type of plate boundary is shown in the diagram? Give an example of where this type of boundary can be found on Earth.

5. Base your answers to the questions on the picture below and on your knowledge of science. The picture shows some high peaks in the Alps in Switzerland.

(a) In what stage of development are the Alps?

(b) What agent of erosion is shown in the picture?

(c) The tectonic forces affecting the Alps are greater than the erosion forces. If this continues, how will the Alps change in the future?

Chapter Puzzle (*Hint:* The words in this puzzle are terms used in the chapter.)

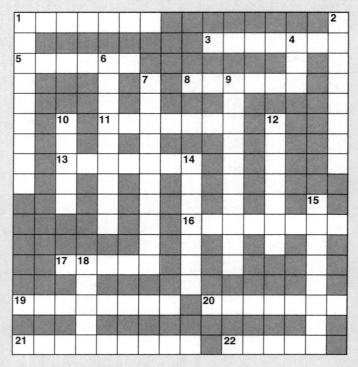

Across

1 a series of large ocean waves caused by an earthquake

3 an opening in Earth's surface that releases lava

5 depression at the top of a volcano

8 deep, elongated gorge on the ocean floor

11 _____ lines connect points of equal elevation

13 coarse rock particles produced by a volcanic eruption

16 _____ spreading refers to the ocean floor moving like a conveyor belt

17 theory of continental _____ suggests the continents are moving

19 large land feature defined by height, slope, and rock type

20 large, generally flat landform at a high elevation

21 crustal plates moving apart

22 large, generally flat landform at a low elevation

Down

1 theory of plate _____ proposes that Earth's crust consists of moving plates

2 area where crustal plates make contact

4 fine dust particles produced by a volcanic eruption

6 point on surface directly above where earthquake occurs

7 crustal plates coming together

9 a sudden shaking in Earth's crust

10 underground location where earthquake occurs

12 side view, or cross section of a cutaway section of Earth's surface

14 vibration waves produced by an earthquake

15 landform significantly higher than the surrounding land

18 mid-ocean_____ is a chain of mountains on the ocean floor

Unit 2

Interactions Between Matter and Energy

Part 1—Essential Question

This unit focuses on the following essential question:

How do the properties and interactions of matter and energy explain physical and chemical change?

Everything in the universe consists of either matter or energy. Matter is anything that has mass and takes up space. All the physical things around us are composed of matter. The water in the ocean, the air we breathe, and the rock beneath our feet are all different forms of matter. All matter is composed of elements, the building blocks of all substances. Elements are made up of

tiny particles called atoms, which contain protons, neutrons, and electrons. Elements may combine to form compounds.

Energy is the ability to do work. That is, energy is the ability to move an object over a distance. There are six different forms of energy: mechanical, chemical, nuclear, heat, electrical, and radiant energy. Last year you learned about mechanical and heat energy, and the transformation of energy from one form into another. This year you will learn about sound, a type of mechanical energy, and light, a type of radiant energy. The properties of sound and light differ in some ways and are similar in other ways.

Matter may change physically or chemically. Both of these changes involve energy. A physical change does not change the identity of the substance, while a chemical change does. A physical change occurs when heat is used to melt a solid or boil a liquid. A chemical change occurs when gasoline burns to release energy, and forms carbon dioxide and water. Photosynthesis in green plants is a chemical change that uses light energy from the sun to change carbon dioxide and water into glucose, a sugar.

Part II—Chapter Overview

Chapter 5 describes both physical and chemical properties of substances. There are two types of substances, elements and compounds. Elements are the simplest kinds of matter. Elements can be combined to form compounds. Each element and compound has a unique structure and specific physical and chemical properties. You will learn about a chart called the periodic table that organizes the elements.

Chapter 6 focuses on two forms of energy – light and sound. Light is a type of radiant energy that can travel through empty space. Because it contains an electric and magnetic field, it is also called electromagnetic energy. There are other types of electromagnetic energy, but light is the only one we can see. Sound is a type of mechanical energy that must travel through a substance, such as air or water, as a series of compressions and expansions. Some properties of light and sound are described and compared.

Chapter 7 describes the characteristics of physical and chemical changes. Dissolving is one kind of physical change. When substances dissolve, they form a special kind of mixture called a solution. This chapter shows how solutions are made and how they are separated. You will learn the role of temperature in the dissolving process. You will see how graphs are used to show the relationship between temperature and solubility.

Chapter 8 discusses chemical reactions, the Law of Conservation of Mass, and the Law of Conservation of Energy. You will learn how to balance chemical equations so that they obey the Law of Conservation of Mass. The types of energy changes that occur in chemical reactions are also described. The role of energy in chemical changes is illustrated using two processes found in nature—photosynthesis and respiration.

Chapter 5

Chemistry: Properties of Matter

Contents

A diamond is the hardest naturally occurring substance. It is made entirely of carbon atoms.

What Is This Chapter About?

Chemistry is the study of the properties, changes, and energy of matter.

In this chapter you will learn:

1. Physical and chemical properties can be used to identify matter.

2. Elements and compounds are two different types of matter.

3. Elements are made up of atoms that are represented by symbols.

4. The structure of an atom helps determine the properties of the element.

5. Elements combine to form compounds that are represented by formulas.

6. Elements are organized into the periodic table based on their properties.

Science in Everyday Life: Chemistry

You probably think of a metal as a shiny solid. You may also be able to identify some of the many different metals that are found in your home. Metals such as gold and silver are often used for jewelry. Copper is used for electrical wires, and aluminum is used to make soda cans and the food wrap known as aluminum foil. How each metal is used is determined by the particular properties of that metal. Chemists study the properties of materials to help them determine which materials are best suited for a particular job.

Internet Sites:

You will find interesting information and activities on chemistry at the following Web sites:

http://www.chem4kids.com An introduction to the topics discussed this year and many more.

http://www.webelements.com This site has an interactive periodic table.

5.1 How Do We Identify Matter?

Objectives

Distinguish between physical and chemical properties.

List several properties that are used to identify matter.

Terms

matter: anything that has mass and takes up space

property (PRAHP-uhr-tee): a characteristic used to describe matter

substance: a pure form of matter with one set of properties

physical property: a property of a substance that can be observed (such as color) and that does **not** involve the formation of a new substance

chemical property: a property that involves the formation of a new substance

flammability (flah-muh-BILL-ih-tee): the ability to burn

malleable (MAL-ee-uh-bull): able to be hammered into shapes

luster: shininess

What Is Matter?

Chemistry is the study of matter, the way matter changes, and the energy that accompanies those changes. **Matter** is defined as anything that has mass and takes up space. All solids (such as salt and sugar), liquids (such as water and gasoline), and gases (such as oxygen and carbon dioxide) are made up of matter.

Physical Properties

Have you ever mistaken salt for sugar? **Stop and Think**. How are salt and sugar similar? How might you tell them apart? **Answer**. Of course both salt and sugar are matter. Both

are also white solids that dissolve in water. However, salt and sugar have very different tastes. Chemists, faced with similar problems, identify matter by examining its **properties**. A **substance** is a pure form of matter with one set of properties. Salt and sugar are both substances. But they have different properties.

A difference in taste helps you distinguish salt from sugar. A difference in color (as well as taste) helps you distinguish salt from pepper. Taste and color are **physical properties**—properties of a substance that are characteristic of that substance and do not involve the formation of a new substance. Testing a physical property does not change the identity of a substance. It is dangerous to

Table 5.1-1. Examples of Physical Properties of Substances

Property	Examples
Phase	Mercury is a liquid at room temperature. Oxygen is a gas at room temperature.
Color	Sulfur is yellow. Chlorine is green.
Odor	Hydrogen sulfide smells like rotten eggs.
Conductivity	Copper is a good conductor of electricity. Glass is a poor conductor of electricity.
Density	Lead is much denser than aluminum. Aluminum is denser than water.
Melting point	Ice melts at 0°C. Table salt melts at 801°C.
Boiling point	Water boils at 100°C. Ethanol (alcohol) boils at 78°C
Solubility	Salt dissolves in water. Gold does not dissolve in water.
Magnetism	Iron is attracted to a magnet. Gold is not attracted to a magnet.

As a reminder: *phase* describes whether a substance is a solid, liquid, or a gas. *Conductivity* measures how easily electricity moves through a material. The *density* of a substance is its mass per unit volume. The *melting point* is the temperature at which a solid turns into a liquid, while the *boiling point* is the temperature at which the liquid turns into a gas. *Solubility* measures the ability of one substance to dissolve in another.

taste a chemical, so chemists never use taste to identify unknown substances. There are, however, many other physical properties that scientists do use to identify matter. Table 5.1-1 lists some of these physical properties.

Chemical Properties

Alcohol can be distinguished from water by its odor and by its boiling point. A chemist can also distinguish alcohol from water by trying to burn it. (See Figure 5.1-1.) Alcohol burns but water does not. Testing the odor or the boiling point does not destroy the alcohol. When you burn alcohol, however, you change the alcohol into new substances (carbon dioxide and water). A property that involves the formation of new substances is called a **chemical property**. **Flammability** is a chemical property that indicates whether or not a substance burns. Table 5.1-2 on page 142 lists several chemical properties.

Figure 5.1-1. Alcohol burns.

Table 5.1-2   Examples of Chemical Properties of Substances

Property	Example
Flammability	The gas in an oven burns
Acidity	Lemon juice is an acid
Reactivity with water	The metal sodium explodes when placed in water
Reactivity with air	Iron rusts
Reactivity with acid	The mineral limestone bubbles when placed in acid

Properties of Gold and Silver

Gold and silver are metals that are frequently used to make jewelry. These metals have certain physical and chemical properties that make them suitable materials for bracelets, rings, and necklaces. Both metals are **malleable**. A malleable substance is one that can be hammered into shapes without breaking. Gold and silver are also both shiny. Chemists call this shininess **luster**. Luster and malleability are both physical properties.

Silver and gold also share some important chemical properties as well that also make them suitable for jewelry. You do not have to worry about getting them wet. Neither metal reacts with water. Some people develop a green stain on their skin when they wear copper jewelry because copper may react with perspiration, but real gold and silver never stain the skin.

What properties would you use to distinguish silver from gold? Let's first consider their physical properties. The most obvious difference is in their color. Silver and gold also differ in density and melting point. They differ very little in odor, solubility, and conductivity, and neither metal is attracted to a magnet.

One chemical property that distinguishes silver from gold is the way silver changes over time. Have you ever needed to polish silver jewelry? Silver forms a black coating, called tarnish, when it reacts with chemicals in the air. Gold does not tarnish. The reactivity of gold is much less than that of silver. This chemical property makes gold very valuable. Gold jewelry will last thousands of years without any change in its appearance.

Activity

Each numbered item below lists some properties of a substance found at home or in school. Use the properties to identify each substance.

1. Solid, soluble in warm water, slippery to the touch, dissolves oils, and removes dirt from surfaces
2. Colorless or pale yellow liquid, insoluble in water, floats on the surface of water, high boiling point
3. White solid, rubs off easily when rubbed against other solids, insoluble in water, does not burn

Interesting Facts About Gold

Gold is so valuable that people often try to sell items that they think are made of gold to a jeweler. How does the jeweler determine whether the items are truly gold? Gold can be distinguished from other metals by its density, but the density of small objects can be difficult to measure accurately. Instead, the jeweler may use a simple chemical test. Nearly all other metals react with a chemical called nitric acid. Because this is a chemical test, it destroys the material being tested. A penny placed in nitric acid will completely disappear! How does the jeweler test the item without destroying it? The jeweler removes just a tiny amount of the metal by scratching the edge of the item on a hard stone plate. Then a small amount of nitric acid is added to the scratch on the plate. If there is no change in the scratch, the sample is genuine gold. People often refer to a test to see if a statement or idea is genuine as an "acid test." This expression originated during the American gold rush, when miners were trying to determine whether the yellow substance they had found was really gold.

Questions

1. Which of the following is a chemical property?
   (1) density
   (2) odor
   (3) flammability
   (4) conductivity

2. Iron is a magnetic, malleable, gray solid that reacts with oxygen to form rust. Which of these properties of iron is a *chemical* property?
   (1) malleability
   (2) color
   (3) phase
   (4) reactivity with oxygen

3. Which of the following describes a physical property?
   (1) silver tarnishes in air
   (2) alcohol burns
   (3) sugar dissolves in water
   (4) gold does not react with water

Thinking and Analyzing

1. A scientist observes the following properties of calcium carbonate:
   A. white solid
   B. reacts with acids to produce carbon dioxide
   C. does not burn
   D. does not dissolve in water
   Which of these properties are physical properties and which are chemical properties?

2. What is the difference between a physical property and a chemical property?

What Are the Two Types of Substances?

Objectives

Define elements.

Define compounds.

Show how symbols are used to represent elements.

Show how formulas are used to represent compounds.

Terms

element: a substance that cannot be broken down into a simpler substance

atom: the smallest piece of an element that has the properties of that element

compound: a substance containing two or more different elements

chemical formula: a way of representing a substance using symbols and numbers

subscript: a number, in a formula, written below and to the right of the symbol

---

Elements

All substances can be classified either as elements or as compounds. An **element** may be defined as a substance that cannot be broken down into anything simpler. Gold, silver, iron, chlorine, carbon, hydrogen, and oxygen are all elements. Each element is represented by a symbol made up of one or two letters. For example, the symbol for hydrogen is H, the symbol for chlorine is Cl, and the symbol for carbon is C. The first letter of the symbol is always capitalized, while the second letter, if there is one, is always lowercased. There are more than 110 known elements.

Only 90 elements occur naturally on Earth; the rest are created in research laboratories. More than half of these 90 elements are rarely found in nature. Table 5.2-1 lists the elements most commonly found in Earth's crust.

Atoms

All elements are made up of tiny particles called atoms. An **atom** is the smallest particle of an element that has the properties of that element. The element oxygen, for example, is composed entirely of oxygen atoms. A piece of pure gold is composed entirely of gold atoms. A second way of defining elements is to say that an *element* is a substance that is made up of just one type of atom. (Atoms will be discussed more fully in Lesson 5.3.)

Table 5.2-1. Elements Commonly Found in Earth's Crust

Element	Chemical Symbol
Oxygen	O
Silicon	Si
Aluminum	Al
Iron	Fe
Calcium	Ca
Sodium	Na
Potassium	K
Magnesium	Mg

Compounds

All substances are made up of atoms, but not all substances are elements. You probably know that water is H_2O. Water contains *two* elements, hydrogen and oxygen. This means that water contains two kinds of atoms. A substance that contains two or more different elements is called a **compound**. While there are only about 110 elements, there are millions of known compounds. The atoms in each compound combine in a specific ratio. When we say that water is H_2O, we mean that water contains twice as many hydrogen atoms as oxygen atoms.

Stop and Think: What substance do you get when you combine a metal that explodes when placed in water, with a poisonous green gas? **Hint:** You sprinkle it on your French fries. **Answer:** The explosive metal is the element sodium, and the green gas is the element chlorine. Together, they form the compound sodium chloride, also known as table salt. Each compound has its own unique set of properties. These properties are completely different from those of the elements from which they are made.

Look back at the list of elements found in Earth's crust, in Table 5.2-1. All of these elements are found in *compounds* in Earth's crust, never as pure elements!

Formulas

The expression "H_2O" is an example of a **chemical formula**. A chemical formula uses symbols and numbers to represent a substance. The symbols indicate which elements are present, and the numbers indicate in what ratio they combine. The "2" in H_2O indicates two hydrogen atoms. The number following the symbol is called a **subscript** because it is written below the letters ("sub" means *below,* "script" means *writing.*) The number "1" is never included in a chemical formula. If no number appears, it means "1." Water contains one oxygen atom. Carbon dioxide has the formula CO_2. This formula tells us that there are two oxygen atoms for each carbon atom in carbon dioxide. Table 5.2-2 on page 146 lists some common substances and their chemical formulas. Notice that ethanol, sucrose, and acetic acid contain exactly the same three elements: carbon, hydrogen, and oxygen. *Stop and Think:* Why are these three compounds so different if they are made up of the same elements? **Answer:** The elements are combined in different ratios.

Stop and Think: The element cobalt has the symbol Co. The compound carbon monoxide has the formula CO. How can we tell them apart? **Answer:** Since a compound is made of more than one element, the

Table 5.2-2. Formulas of Common Substances

Substance	Formula	Composition
Methane (natural gas)	CH_4	1 – carbon atom 4 – hydrogen atoms
Ethanol (alcohol)	C_2H_6O	2 – carbon atoms 6 – hydrogen atoms 1 – oxygen atom
Sucrose (table sugar)	$C_{12}H_{22}O_{11}$	12 – carbon atoms 22 – hydrogen atoms 11– oxygen atoms
Acetic acid (vinegar)	$C_2H_4O_2$	2 – carbon atoms 4 – hydrogen atoms 2 – oxygen atoms
Table salt	NaCl	1 – sodium atom 1 – chlorine atom

formula of any compound must contain at least **two** capital letters. The formula for an element must contain only **one** capital letter. Ozone gas has the formula O_3. It is considered an element because it contains only one type of atom, oxygen. Is NO an element? No, but No is! (NO is the formula for the compound nitrogen monoxide. No is the symbol for the man-made element nobelium.)

Activity

Go to the website listed below. Study each chemical formula and determine whether the substance it represents is an element or a compound.

http://www.curriculumbits.com/prodimages/details/chemistry/che0001.html

Interesting Facts About Atoms and Elements

The word "atom" comes from the Greek word "atomos," which means, "Cannot be divided." The idea that matter is composed of tiny atoms was suggested in about 400 B.C. (almost 2,500 years ago). The Greeks had very different ideas about the composition of matter. They thought that all atoms were the same. They also thought that there were just four elements: earth, air, fire, and water. Even today, the word "element" sometimes keeps its ancient meaning. Objects that are left outside, unprotected from rain, wind, or heat, are said to be "exposed to the elements."

Questions

1. Which of the following represents an element?
   (1) H_2O (3) O_2
   (2) CH_4 (4) $C_6H_{12}O_6$

2. The simplest substances are called
   (1) elements   (3) formulas
   (2) compounds   (4) symbols

3. Chemical formulas are used to represent
   (1) elements only
   (2) compounds only
   (3) both elements and compounds
   (4) neither elements nor compounds

4. Which statement is true about Hf and HF?
   (1) Both are elements.
   (2) Both are compounds.
   (3) Both are substances.
   (4) Hf is a compound while HF is an element.

5. Which term includes the other three?
   (1) elements   (3) atoms
   (2) compounds   (4) matter

Thinking and Analyzing

1. How many hydrogen atoms, carbon atoms, and oxygen atoms are represented by the formula H_2CO_3?

2. When the mineral bauxite is melted and exposed to an electric current, it produces two new substances, one of which is aluminum. Is bauxite an element or a compound? Justify your answer.

3. Some common substances and their chemical formulas are listed in the table below.
   Copy this table in your notebook. Indicate which of these substances are elements and which are compounds by checking the appropriate column. The first one has been done for you.

Substance	Formula	Element	Compound
Carbonic acid	H_2CO_3		√
Nitrogen	N_2		
Hydrochloric acid	HCl		
Carbon dioxide	CO_2		
Water	H_2O		
Helium	He		

5.3 What Are the Parts of an Atom?

Objectives

Identify three subatomic particles based on charge, mass, and location.

Draw and use a model to represent an atom.

Use the atomic number to determine the number of protons and electrons in an atom.

Terms

subatomic particle: a particle that is smaller than an atom

proton: a positively charged subatomic particle found in the nucleus

neutron (NU-tron): a neutral subatomic particle found in the nucleus

electron: a negatively charged subatomic particle found outside the nucleus

nucleus (NU-klee-uhs): the center of an atom containing the neutrons and protons

atomic number: the number of protons in an atom

Atoms

Imagine taking a piece of an element, such as aluminum, and breaking it into smaller and smaller pieces. This smallest piece of aluminum that you could make is called an aluminum atom. If you break this piece any further, it will no longer be aluminum. An atom is so small that it cannot be seen even with the most powerful microscopes. A single aluminum atom has a radius of 0.000000000143 (1.43×10^{-10}) meter. This means that if you could place aluminum atoms end to end, a centimeter would contain about 140 million aluminum atoms. Do not try to memorize these numbers! Just remember that atoms are *really, really, really* small.

Subatomic Particles

As small as atoms are, they are made up of particles that are even smaller. These are called **subatomic particles**. Subatomic particles include **protons**, **neutrons**, and **electrons**. These particles differ in mass, electrical charge, and location in the atom. Protons and neutrons have roughly the same mass, while electrons are much lighter. Protons have a positive (+) charge, and electrons have a negative (−) charge. Neutrons have no electrical charge; they are electrically neutral. Protons and neutrons are found in the center, or **nucleus**, of the atom. Electrons orbit the nucleus, moving very rapidly. Because the mass of an electron is much smaller than that of a proton or a

Table 5.3-1. Properties of the Subatomic Particles

Particle	Mass (AMU)	Charge	Location	Abbreviation
Proton	1	+	Nucleus	p^+
Neutron	1	0	Nucleus	n^0
Electron	0.00054	−	Outside the nucleus	e^-

neutron, nearly all of the mass of an atom is found in the nucleus. Table 5.3-1 summarizes the properties of the subatomic particles.

Opposites Attract

All charged particles are either positively charged or negatively charged. They obey the simple rule that like charges repel, while opposite charges attract. (See Figure 5.3-1.) The negatively charged electrons in an atom are attracted to the positively charged protons. This attraction prevents the electrons from flying off the atom.

Recall that elements are made up of one type of atom. Different elements contain different types of atoms. How do these atoms differ? Atoms of different elements have different numbers of protons in their nucleus. Oxygen has 8 protons, while carbon has 6, and uranium has 92. The number of protons in the nucleus is called the **atomic number**. All atoms of the same element have the same atomic number and therefore the same number of protons.

The mass of an atom is measured in atomic mass units (AMU), a special unit created for measuring the mass of very small particles. You can see from Table 5.3-1 that almost all of the mass of an atom is found in the nucleus. The approximate mass of an atom, in AMU, can be found by adding the number of protons to the number of neutrons. For example, if an oxygen atom has 8 protons and 8 neutrons, its mass is 16 AMU.

If you know the atomic number and the mass of an atom, you can calculate the number of neutrons. For example, consider an atom of fluorine with a mass of 19 AMU. The atomic number of fluorine is 9. *Stop and Think:* How many neutrons are there in this fluorine atom? *Answer:* The total

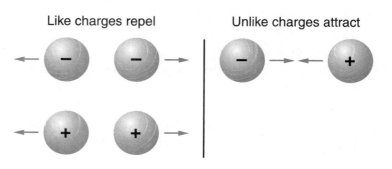

Like charges repel   Unlike charges attract

Figure 5.3-1. Two charges that are the same are called like charges. Like charges repel each other. Opposite charges (+ with −) are called unlike charges. Unlike charges attract each other.

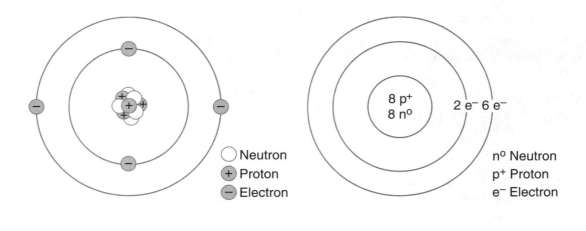

(a) A beryllium atom   (b) An oxygen atom

Figure 5.3-2.   Two models of an atom. (a) A beryllium atom has 4 protons, 5 neutrons, and 4 electrons. (b) An oxygen atom has 8 protons, 8 neutrons, and 8 electrons.

number of protons and neutrons is 19. Since the atomic number is 9, we know there are 9 protons. Therefore, there must be 10 neutrons in this atom.

$$\begin{array}{r} 19 \text{ protons} + \text{neutrons} \\ -9 \text{ protons} \\ \hline 10 \text{ neutrons} \end{array}$$

Scientists represent atoms with various models. In these models, the protons and neutrons are drawn in the center, and the electrons are drawn surrounding the nucleus. Figure 5.3-2(a) shows a model of a beryllium atom, with the protons, neutrons, and electrons drawn in. Figure 5.3-2(b) shows a model of an oxygen atom by indicating the number and location of the subatomic particles.

Notice that in Figure 5.3-2, the electrons are distributed around the nucleus in two separate rings. Electrons do not really move

in circular orbits. These rings represent locations that modern chemists call "shells" or "energy levels." Each shell can hold a limited number of electrons. Once one shell is filled, a new one begins. The maximum number of electrons in the first shell is 2. The second shell can hold up to 8 electrons.

Neutral Atoms

All atoms are electrically neutral. *Stop and Think:* How can atoms be neutral if they contain particles that have positive and negative charges? *Answer:* The total positive charge in an atom equals the total negative charge in that atom. In order for an atom to be neutral, the number of positive protons must be equal to the number of negative electrons. In Figure 5.3-2(a), the beryllium atom has 4 protons and 4 electrons. In Figure 5.3-2(b), the oxygen atom has 8 protons and 8 electrons. Because they have no charge, the number of neutrons does not have to be equal to the number of protons or electrons.

A sodium atom has an atomic number of 11, and contains 12 neutrons. What might a model of this atom look like? Since the atomic number is 11, we know that sodium must contain 11 protons. Since the number of protons in an atom always equals the number of electrons, sodium must contain 11 electrons. Because the first two shells can hold only 10 electrons (2 in the first shell and 8 in the second), we need

a third shell for the last electron. Figure 5.3-3 shows a model of a sodium atom.

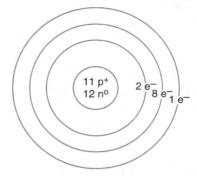

Figure 5.3-3. A model of a sodium atom.

Activity

Visit *http://education.jlab.org/atomtour/* on the Internet and view the activity. Answer the following questions:

1. What is the name of the particle that protons and neutrons are made of?
2. How many electrons does it take to equal the mass of one proton?
3. What do we call a heavier version of an atom of the same element?

Interesting Facts About Atomic Structure

The model of the atom is constantly changing as new information becomes available. The original concept of an atom was that it was the smallest possible particle of matter. When electrons were discovered around 1900, that theory had to change. The later discoveries of the nucleus, proton, and neutron led to new models of the atom.

Chemists also once thought that atoms were indivisible (cannot be divided). Scientists also discovered that this is not true. When a neutron strikes a certain kind of uranium nucleus, the nucleus splits into two smaller atoms. This process, called nuclear fission (pronounced NU-klee-uhr, NOT NU-kyu-luhr), also produces tremendous amounts of energy. This energy is used in nuclear power plants, like the Indian Point reactor in New York, to produce electricity. The energy from fission is also used in the powerful weapons known as atomic bombs.

Questions

1. Which statement is true of all atoms?
   (1) The number of protons equals the number of electrons.
   (2) The number of neutrons equals the number of electrons.
   (3) The number of protons is less than the number of electrons.
   (4) The number of protons is greater than the number of neutrons.

2. The nucleus of the atom contains
   (1) protons and electrons
   (2) protons and neutrons
   (3) neutrons and electrons
   (4) protons, neutrons, and electrons

3. Which particle is correctly matched with its charge?
   (1) proton—neutral
   (2) electron—positive
   (3) neutron—negative
   (4) electron—negative

The element nitrogen has an atomic number of 7. Use this information to answer questions 4–6.

4. How many electrons are found in an atom of this element?
   (1) 14   (3) 5
   (2) 2   (4) 7

5. How many energy levels are there in an atom of nitrogen?
   (1) 1   (3) 3
   (2) 2   (4) 4

6. A nucleus of this atom might contain
   (1) 8 protons and 7 neutrons
   (2) 8 protons and 8 electrons
   (3) 7 protons and 8 neutrons
   (4) 7 protons and 7 electrons

7. Which chemical principle is illustrated by the cartoon below?
   (1) like charges attract
   (2) like charges repel
   (3) opposite charges attract
   (4) opposite charges repel

Thinking and Analyzing

1. Aluminum has an atomic number of 13. Using the model illustrated in Figure 5.3-3, draw a model of an aluminum atom having 14 neutrons.

2. Use Table 5.3-1 on page 149 to determine the mass of your aluminum atom to the nearest whole number.

3. Why are atoms electrically neutral?

Base your answers to questions 4–7 on the model of the atom below.

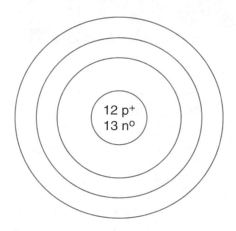

4. What is the atomic number of this atom?

5. What is the mass of this atom?

6. What subatomic particle is left out of the model?

7. Copy the atom into your notebook and complete it by adding the missing parts.

5.4

How Do Atoms Combine?

Objectives

Define molecules.

Describe two types of chemical bonds.

Interpret models of molecules.

Terms

molecule: a particle that contains two or more atoms bonded together

chemical bond: an attraction that holds two atoms together

structural formula: a model that shows the arrangement of atoms in a molecule

covalent (koh-VAY-luhnt) **bond:** a chemical bond in which electrons are shared

ion (EYE-on): a positively or negatively charged particle formed by the loss or gain of electrons

ionic (eye-ON-ick) **bond:** a chemical bond formed by the attraction of oppositely charged ions

Molecules

Recall that the smallest particle of an element is called an *atom*. The smallest particle of a compound that still has the properties of that compound is called a **molecule**. A molecule is a particle containing two or more atoms bonded together. A **chemical bond** is an attraction that holds atoms together. You know that the formula for water is H_2O. A water molecule contains two hydrogen atoms and one oxygen atom. Figure 5.4-1 shows one model used to represent a water molecule. This type of model shows the shape, or *geometry*, of the molecule.

A *chemical formula* tells us exactly what the molecule contains. Formulas are often written to show not only which atoms are in

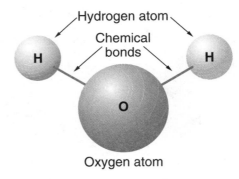

Hydrogen atom

Chemical bonds

H

H

O

Oxygen atom

Figure 5.4-1. A water molecule.

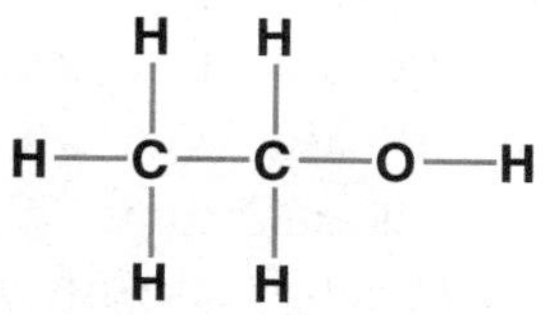

H   H
H—C—C—O—H
H   H

Figure 5.4-2. The structural formula of ethanol shows which atoms are bonded to which other atoms.

the molecule, but also how they are arranged. In Table 5.2-2 on page 146, the formula for ethanol is listed as C_2H_6O. Figure 5.4-2 shows the arrangement of the atoms in a molecule of ethanol.

To better represent this molecule, chemists usually write the formula for ethanol as C_2H_5OH. Models like those shown in Figures 5.4-1 and 5.4-2 are called **structural formulas**. Acetic acid, listed in Table 5.2-2 on page 146 as $C_2H_4O_2$, has the structure shown in Figure 5.4-3. To illustrate this arrangement, chemists generally write the formula as CH_3COOH.

Chemical Bonds

The atoms in a molecule are held together by chemical bonds. The lines that connect the atoms in Figures 5.4-1 and 5.4-2 represent these bonds. The bonds shown in these diagrams are called **covalent bonds**. Covalent bonds are formed when atoms

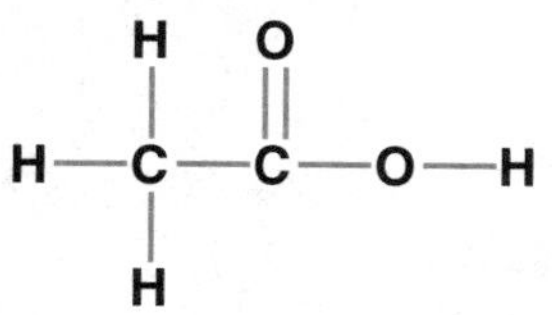

H   O
H—C—C—O—H
H

Figure 5.4-3. The structural formula for acetic acid.

share electrons. In the water molecule, each hydrogen atom shares two electrons with the oxygen atom. Each line connecting a hydrogen atom to the oxygen atom represents two shared electrons. A covalent bond always involves the sharing of a pair of electrons. Look at the structure of acetic acid in Figure 5.4-3. **Stop and Think:** What do the two lines, connecting the carbon to the oxygen above it, represent? **Answer:** The two lines represent four shared electrons, or two pairs. We call such a bond a *double bond*. Some molecules even have *triple bonds*!

Ionic Bonds

Recall that table salt, sodium chloride, has the formula NaCl. In this case, the bond between the atoms does not involve sharing electrons. When sodium reacts with chlorine, an electron actually moves from the sodium atom to the chlorine atom. Chlorine atoms have 17 protons and 17 electrons. When chlorine takes an electron from sodium, it ends up with 18 electrons, but it still has 17 protons. The chlorine is no longer neutral! It now has more electrons than protons giving it a negative charge. A sodium atom has 11 protons and 11 electrons. When it loses an electron to the chlorine, it ends up with only 10 electrons. Since it now has more protons than electrons, it has a positive charge. These charged particles are no longer atoms, because all atoms must be neutral. The positive sodium and negative chlorine particles are called **ions**. An **ion** is a positively or negatively charged particle.

Remember that opposite charges attract. The positive sodium ion, or Na^+, attracts the negative chloride ion, or Cl^-. They come together to form an ionic bond. An **ionic bond** is the attraction between oppositely charged ions.

You might have noticed that the name of the positive ion is the same as the name of the atom, sodium, while the name of the negative chloride ion is *different* from the name of the chlorine atom. The names of negative ions are always different from the names of the neutral atoms. When the elements magnesium and sulfur combine, for example, they form magnesium sulfide.

Interesting Facts About Molecules

The smallest naturally occurring molecule is a molecule of hydrogen gas, H_2. This molecule contains just two of the smallest possible atoms. One hydrogen molecule weighs 0.000000000000000000000000332 gram.

What is the largest naturally occurring molecule? A diamond crystal is considered a single covalently bonded molecule containing only carbon atoms. (See Figure 5.4-4.) The largest one found to date was named the Cullinan, with a mass of 621.2 grams, or 3106 carats. For information on the Cullinan diamond, visit

http://www.diamondvues.com/2005/03/the_history_of.html

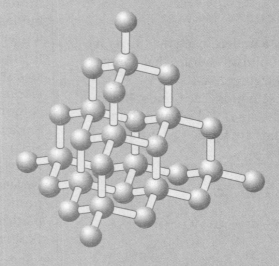

Figure 5.4-4. The structure of a diamond crystal. Each ball represents a carbon atom, and each stick represents a covalent bond.

Activity

Visit this Web site to get more information about chemical bonds:
http://www.visionlearning.com/library/module_viewer.php?mid=55

Read up to "The reaction of sodium with chlorine." Then click on the link to watch the flash video. When you see the video, click on the directions in the bottom right-hand corner of the window:

1. Drop the sodium into the chlorine gas.
2. Magnify the reaction.

Answer these questions:

3. What compound is formed?
4. Describe what happens when sodium is dropped into chlorine.

Questions

1. The smallest particle of a compound that still has the properties of that compound is called a(n)
   (1) atom   (3) neutron
   (2) element   (4) molecule

2. Atoms that share electrons form a(n)
   (1) ionic bond   (3) element
   (2) covalent bond (4) nucleus

3. The formula H_2SO_4 indicates that each molecule of this compound contains
   (1) 1 sulfur atom
   (2) 2 sulfur atoms
   (3) 1 oxygen atom
   (4) 4 hydrogen atoms

4. A particle that contains 9 protons and 10 electrons is best described as
   (1) an atom with a positive charge
   (2) an atom with a negative charge
   (3) an ion with a positive charge
   (4) an ion with a negative charge

Thinking and Analyzing

Refer to the diagram below to answer questions 1–3.

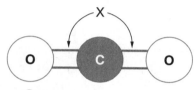

Ⓒ Carbon atom
Ⓞ Oxygen atom

1. What does this diagram represent?
2. What is represented by "X"?
3. What is a possible formula for this compound?
4. Explain the difference between an ionic bond and a covalent bond.

How Are Elements Arranged in the Periodic Table?

Objectives

Find similar elements based on their position in the periodic table.

Identify metals, nonmetals, and noble gases based on their position in the periodic table.

Identify metals, nonmetals, and noble gases based on their properties.

Terms

group: a column in the periodic table of the elements

valence (VAY-lentz) **electrons:** electrons found in the outermost shell of an atom

valence shell: the outermost shell of an atom

metals: shiny, malleable elements that conduct electricity; metals are found on the left side of the periodic table

nonmetals: elements found on the right side of the periodic table that are poor conductors of electricity

ductile (DUCK-tile): can be stretched into wires

brittle: breaks when hit with a hammer

noble gas: elements found in group 18 of the periodic table that are nonreactive

The Periodic Table of the Elements

Remember that all matter is made of elements. Most matter consists of elements that have combined together to form compounds. To help understand the behavior of these elements, scientists have organized them into a chart called the periodic table of the elements. (See Figure 5.5-1.) The periodic table generally gives the symbol of each element and lists its atomic number and its mass. In this periodic table, the states of the elements are indicated by the color of the symbol.

Groups

In the early 1800s, scientists noticed that certain elements greatly resemble each other in their physical and chemical properties. For example, copper (Cu), silver (Ag), and gold (Au) are all shiny solids that conduct electricity and do not react with water. Sodium (Na) and potassium (K), on the other hand, are shiny solids that explode when placed in water. Find these elements in the periodic table in Figure 5.5-1. What do you notice about their locations? Copper, silver, and gold are found in the same

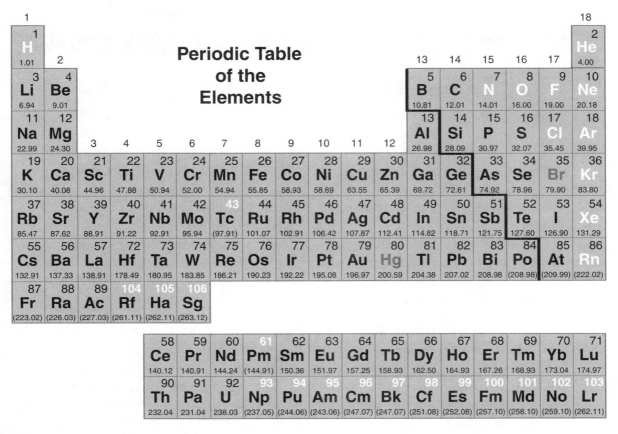

Figure 5.5-1. The periodic table of the elements. White elements are gases, green elements are liquids, and **black** elements are solids at room temperature.

column. Sodium and potassium are found in a different column. In the 1800s, scientists created a list of all the known elements in order of increasing mass. Similar elements were grouped together in the same column. This list became known as the periodic table. The modern periodic table lists the elements in order of increasing atomic number. It is organized into 18 columns called **groups** or families. They are called families because the elements in them resemble each other. The group number is listed on the chart at the top of each column.

The Periodic Law

Why is the table called the "periodic" table? Dmitri Mendeleev, a Russian chemist, studied the properties of the 63 elements that were known in his time (around 1860). He discovered that when he listed these elements in order of their masses, certain properties recurred periodically. For example, if you begin with lithium (Li), you find that 8 elements later you come to sodium, an element that is very similar to lithium. Count another 8, and you get to potassium, which greatly resembles both sodium and lithium. Continue along the

Table 5.5-1.   The Electron Configurations of Some Elements

Group 1		Group 2		Group 16		Group 17	
Li	2-1	Be	2-2	O	2-6	F	2-7
Na	2-8-1	Mg	2-8-2	S	2-8-6	Cl	2-8-7
K	2-8-8-1	Ca	2-8-8-2	Se	2-8-18-6	Br	2-8-18-7

table, and every so often you will come to another element that resembles lithium. The same physical and chemical properties recur periodically.

Stop and Think: Mendeleev listed the elements in order of mass. Modern chemists list them in order of atomic number. Does it make a difference? **Answer:** The table looks *almost* the same whether we use mass or atomic number. One difference, however, is that a table based on mass would place element 52, Te, with a mass of 127.6 *after* element 53, iodine, with a mass of 126.9. Mendeleev's table does not include the atomic number because he developed it in 1869, and the atomic number was not discovered until 1913.

Chemists summarize the relationships among the elements through the periodic law. The periodic law states that when elements are listed in order of their atomic numbers, similar physical and chemical properties will appear periodically. Using the periodic law, Mendeleev was able to predict the properties of elements that had not yet been discovered.

The Valence Shell

Mendeleev did not know why the periodic law worked as well as it did. We now know that the properties of each element depend largely on the arrangement of its electrons.

Figure 5.3-3 on page 151 shows the arrangement of the electrons in a sodium atom. Notice that the outermost shell contains one electron. All of the other elements in the same group as sodium also contain one outermost electron. Chemists say that the electron configuration of sodium is 2-8-1. This means that sodium has 2 electrons in its first shell, 8 in its second, and 1 in its third, or outermost shell. Table 5.5-1 lists the electron configurations of several elements.

Stop and Think: What do you notice about the elements in these groups? **Answer:** The number of electrons in the outermost shell is the same for each element within the group. The outermost electrons are most important in determining the behavior of the element. These electrons are known as the **valence electrons**, and the outermost shell is called the **valence shell**. You may also notice that as you go down each group, the number of electron shells increases.

Metals and Nonmetals

You are probably familiar with several pure elements. Iron (Fe), tin (Sn), aluminum (Al), nickel (Ni), copper (Cu), silver (Ag), and gold (Au) are all shiny solids that conduct electricity. Elements that share these properties are called **metals**. Most elements are metals. *Stop and Think:* Can you think of an element that is not a metal? ***Answer:*** Oxygen (O) and nitrogen (N) are invisible gases and are not metals. We call them **nonmetals**. Other nonmetals include carbon (C), sulfur (S), and iodine (I).

Look at the periodic table on page 159 and locate each of the nonmetals listed above. Compare the locations of these nonmetals with those of the metals. You can see that in general, nonmetals are on the right side of the table, and metals on the left. There is a dark line forming what looks like a staircase separating the metals from the nonmetals. Notice that there are many more metals than nonmetals.

Properties of Metals

We have already learned that metals are shiny solids that conduct electricity. (Mercury, symbol Hg, is a liquid at room temperature, but it is shiny and it does conduct electricity, so it is still considered a metal.) Metals share some additional properties. Metals are malleable—when you hit a metal with a hammer, it will bend, not break. Metals are also **ductile**—they can also be stretched into wires. The electricity in your home passes through hundreds of feet of copper wire.

The properties of metals are determined by the arrangement of the electrons in their atoms. Metals usually have only one or two valence electrons. Metals react by losing electrons to form positive ions. They generally form ionic bonds.

Properties of Nonmetals

Nonmetals are located to the right of the "staircase line." Some are solids, some are gases, and one, bromine (Br), is a liquid at room temperature. They come in a variety of colors: sulfur is yellow, bromine is red, iodine is purple, chlorine is green, and phosphorus can be white or red. Others, such as oxygen and nitrogen gas, are colorless. Nonmetallic solids are poor conductors of electricity and are **brittle**— they break when you hit them with a hammer. Nonmetals react with metals by gaining electrons to form negative ions. Nonmetals also react with other nonmetals to form covalent bonds. All nonmetals have four or more valence electrons.

Noble Gases

Group 18, on the right side of the periodic table, contains gases that do not gain or lose electrons. These elements are called the **noble gases**. They are the least reactive of all the elements. No compound has ever been made containing helium, neon, or argon.

Activity

Adopt an element. On what day of the month were you born? Take that day and adopt the element with that atomic number. If you were born on June 8, for example, you will be adopting the element oxygen because its atomic number is 8. Find out about your element. What is it used for? What does it look like? Draw a model of its atomic structure. How was it discovered? Where does its name come from? Start by visiting *http://www.webelements.com/* and clicking on the element's symbol. You can also search for your element on *http://en.wikipedia.org*

Questions

1. Elements in the modern periodic table of the elements are listed in order of
   (1) mass
   (2) size
   (3) color
   (4) atomic number

2. An element that is a good conductor of electricity, solid at room temperature, and can be hammered into shapes is best described as
   (1) a metal
   (2) a nonmetal
   (3) a noble gas
   (4) a molecule

3. Which of the following elements is a nonmetal?
   (1) Na
   (2) Al
   (3) Hg
   (4) S

4. Which of the following elements is a gas?
   (1) Ne
   (2) Fe
   (3) Hg
   (4) C

5. The largest number of elements are
   (1) metals
   (2) nonmetals
   (3) noble gases
   (4) liquids

6. The diagram below shows an *incomplete* circuit.
   Which item would allow the bulb to light up if it were used to connect point A to point B?
   (1) a glass rod
   (2) a metal coin
   (3) a plastic comb
   (4) a paper cup

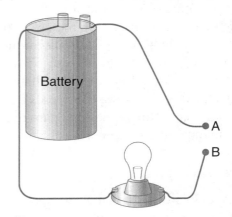

7. Which element would you expect to be most similar to calcium (Ca)?
   (1) K
   (2) Sc
   (3) Mg
   (4) Ar

Thinking and Analyzing

1. The human body uses calcium to build bones and teeth. If strontium is found in the environment, it can be absorbed by the body and used instead of calcium. Strontium can be found in the bones and teeth of people who are exposed to it. Explain why the body often mistakes the element strontium for the element calcium.

2. How are the elements bromine and mercury similar? How are they different?

Review Questions

Term Identification

Each question below shows two terms from Chapter 5. One of the terms is defined.
(1) Choose the term that matches the definition.
(2) Describe how the two terms are different. Following each term is the section (in parenthesis) where the description or definition of that term is found.

1. *Chemical property (5.1) — Physical property (5.1)*
   A property that involves the formation of a new substance

2. *Flammable (5.1) — Malleable (5.1)*
   Able to be hammered into shapes

3. *Element (5.2) — Compound (5.2)*
   A substance that cannot be broken down into a simpler substance

4. *Atom (5.2) — Ion (5.4)*
   The smallest piece of an element that has the properties of that element

5. *Proton (5.3) — Electron (5.3)*
   A positively charged, subatomic particle found in the nucleus

6. *Nucleus (5.3) — Neutron (5.3)*
   The center of an atom

7. *Brittle (5.4) — Malleable (5.1)*
   Breaks when hit with a hammer

8. *Ionic bond (5.4) — Covalent bond (5.4)*
   A chemical bond in which electrons are shared

9. *Valence shell (5.4) — Group (5.4)*
   A column on the periodic table of the elements

10. *Metal (5.5) — Noble gas (5.5)*
    Elements found in group 18 of the periodic table that are nonreactive

Multiple Choice (Part 1)

Choose the response that best completes the sentence or answers the question.

1. The experiment below is being used to determine whether the screwdriver is made of metal. What property is being tested?

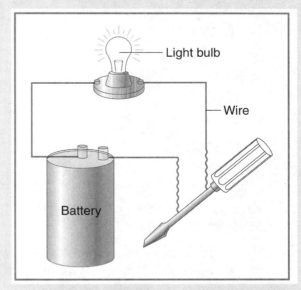

(1) magnetic attraction   (3) malleability
(2) electrical conductivity   (4) flammability

2. The table below shows the chemical symbols for some common elements.

Element	Symbol
Hydrogen	H
Helium	He
Oxygen	O
Silicon	Si
Carbon	C
Iron	Fe

Based on the information in the table, which of the four substances below is a compound?

(1) CO   (3) Si
(2) He   (4) Fe

3. A chemical property of a mineral is evident if the mineral
(1) breaks easily when struck with a hammer
(2) bubbles when acid is placed on it
(3) is easily scratched by a fingernail
(4) reflects light from its surface

4. The diagram below shows the geometric structure of a molecule of water (H_2O).

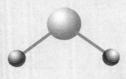

What do the symbols and represent in the model?

(1) genetic material   (3) cells
(2) chemical bonds   (4) atoms

5. Mercury is a silver-colored liquid at room temperature. This statement describes
(1) one chemical property and one physical property
(2) two chemical properties
(3) two physical properties
(4) one chemical property but no physical properties

6. Which is the formula of an element?
(1) Hf   (3) CO_2
(2) HF   (4) H_2O

7. The smallest particle of the element sodium that still has the properties of sodium is called a sodium
(1) electron   (3) atom
(2) nucleus   (4) neutron

8. A neutron walks into a soda shop and orders a drink. It says to the waiter, "How much?" The waiter answers
(1) "For you, no charge."
(2) "Are you positive you do not want a refill?"
(3) "Why the negative attitude?"
(4) "I'm positive I am attracted to you."

9. Based on the periodic table on page 159, the element that contains 9 protons is classified as
(1) a metal   (3) a noble gas
(2) a nonmetal   (4) a metalloid

10. Which particles are found in the nucleus of an atom?
(1) protons and electrons
(2) protons only
(3) protons and neutrons
(4) neutrons only

Base your answers to questions 11 and 12 on the following reading passage and on your knowledge of chemistry.

When sodium is heated in chlorine gas, it bursts into flames and forms a new substance called sodium chloride (NaCl). During this process, electrons are transferred from the sodium atoms to the chlorine atoms.

11. The sodium chloride molecule is held together by an attraction called
(1) a covalent bond
(2) an ionic bond
(3) a nuclear bond
(4) a savings bond

12. This reaction is best described as the reaction between
(1) two metals
(2) two nonmetals
(3) a metal and a nonmetal
(4) a metal and a noble gas

13. Which is a property of the element sulfur (S)?
(1) It is malleable.
(2) It is ductile.
(3) It is brittle.
(4) It is a good conductor of electricity.

14. An unknown element is a malleable, silver-gray solid and is a good conductor of electricity. This element most likely belongs to which of the following groups?
(1) 8   (3) 17
(2) 16   (4) 18

Thinking and Analyzing (Part 2)

Base your answers to questions 1–6 on the model of an atom below.

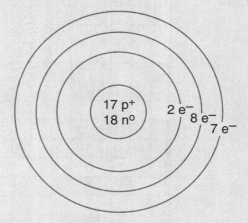

1. What is the symbol for the element that this atom represents?

2. To what group in the periodic table does this element belong?

3. What is the mass of this atom?

4. How many electrons does this atom have in its valence shell?

5. What is the atomic number of this element?

6. What element has properties that are similar to the element in the diagram? What is its symbol?

7. Draw a model of a carbon atom that contains 6 neutrons.

8. When potassium is heated in oxygen, a chemical reaction takes place. When potassium is heated in neon, it does not react. Why doesn't potassium react with neon? (*Note:* symbols for these elements are: potassium – K, oxygen – O, neon – Ne.)

Chapter Puzzle (*Hint:* The words in the puzzle are terms used in the chapter.)

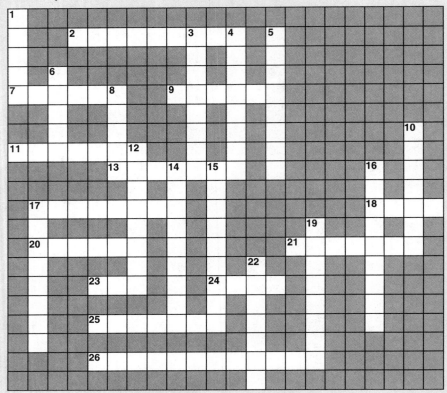

Across

2 a particle that is smaller than an atom is called a _____ particle

7 a positively charged subatomic particle

9 shiny, malleable elements that conduct electricity

11 can be stretched into a wire

13 a substance that cannot be broken down into a simpler substance

17 a property that involves the formation of a new substance is called a _____ property

18 an attraction that holds two or more atoms together is called a chemical _____

20 anything that has mass and takes up space

21 breaks when hit with a hammer

23 a positively or negatively charged particle

24 the smallest piece of an element that has the properties of that element

25 the center of the atom, which contains the protons and neutrons

26 ability to burn

Down

1 a column on the periodic table is called a _____

3 a particle that contains two or more atoms bonded together

4 a bond in which electrons are shared is called a _____ bond

5 a property that can be observed without changing the identity of the substance is called a _____ property

6 a chemical bond caused by the attraction of oppositely charged ions is called an _____ bond

8 group 18 of the periodic table contains nonreactive elements called _____ gases

10 the electrons on the outermost shell are called the _____ electrons

12 a negatively charged subatomic particle

14 able to be hammered into shapes

15 elements found on the right side of the periodic table that are poor conductors of electricity

16 a pure form of matter; an element or a compound

17 a substance containing two or more different elements

19 a characteristic used to describe matter

22 a substance is represented using symbols and numbers in what is called a chemical _____

Chapter 6

Properties of Light and Sound

Contents

Sunlight and raindrops produce a rainbow, one of nature's most colorful sights.

What Is This Chapter About?

Radiant energy is a form of energy that travels as electromagnetic waves. Light is a form of electromagnetic energy. Sound is a form of mechanical energy that travels as a different type of wave. In some ways light and sound are similar, and in other ways they differ.

In this chapter you will learn:

1. The electromagnetic spectrum is the orderly arrangement of different types of radiant energy.

2. Each type of electromagnetic radiation differs in wavelength, frequency, and energy.

3. Light is the visible portion of the electromagnetic spectrum.

4. Light waves that strike an object can be reflected, absorbed, or transmitted by the object. Refraction is the bending of light waves.

5. Sound is a form of mechanical energy produced by a vibrating object.

Science In Everyday Life

Electromagnetic waves are all around us. When you are watching TV, talking on a cell phone, cooking in a microwave oven, or sunbathing you are using electromagnetic waves. Although you can see visible light, most electromagnetic waves can only be detected with special equipment.

Internet Sites:

http://www.colorado.edu/physics/2000/index.pl   At Einstein's Legacy explore the operation of x-rays, microwave ovens, and lasers. Use an x-ray machine called a fluoroscope (FLUR-uh-skohp).

http://www.isvr.soton.ac.uk/SPCG/Tutorial/Tutorial/StartCD.htm   This site contains advanced information about sound waves. The Wave Basic section has animations of different types of waves.

http://www.learner.org/teacherslab/science/light/index.html   This is the Science of Light Internet site. It has interactive activities that demonstrate the principles of light.

6.1 What Is Electromagnetic Energy?

Objectives

Define the characteristics of electromagnetic energy.

Identify the parts of an electromagnetic wave.

Describe the different types of electromagnetic energy.

Terms

electromagnetic (ih-lehk-troh-mag-NEH-tihk) **energy:** different types of radiation that move in waves and can travel through space

wavelength: the distance from the top of one wave to the top of the next wave

amplitude: the height of a wave measured vertically from the undisturbed surface to the top of the wave

frequency: the number of waves that pass by a fixed point in a given amount of time

medium: a substance, such as air or water, through which waves travel

electromagnetic spectrum: a continuous band of electromagnetic waves arranged according to wavelength and frequency

Electromagnetic Energy

As you may remember, energy is the ability to do work. Table 6.1-1 lists the six forms of energy. Each form of energy has the ability to move an object (to do work). For example, throw a pebble in a pond and you see the mechanical energy in the moving pebble transfer its energy to the water. Or watch a battery-operated toy car move across the floor by converting the chemical energy in the battery into the mechanical energy of the moving car.

Electromagnetic energy is radiant energy that moves in waves and can travel through space. The waves are called electromagnetic waves because they contain both an electrical field and a magnetic field. The waves are created by electrically charged particles that vibrate rapidly. Types of electromagnetic energy include: radio waves, microwaves, infrared waves, visible light, ultraviolet light, x-rays, and gamma rays.

Wave Properties

As mentioned before, electromagnetic energy travels in the form of waves. In some ways, electromagnetic waves are similar to the waves produced when a pebble is tossed into a pool of calm water.

Table 6.1-1. Different Forms of Energy

Mechanical energy	Energy of a moving object. Sound is one type of mechanical energy, because air molecules move when sound passes through it.
Chemical energy	Energy stored in the bonds of atoms and molecules.
Nuclear energy	Energy stored in the nucleus (center) of an atom.
Heat energy	Energy associated with vibrating molecules.
Electrical energy	Energy produced by the flow of electrons through a substance.
Radiant energy	Energy that moves in waves and can travel through space. Light is a type of radiant energy that is visible to us.

The pebble hitting the water is a source of energy. When it hits the water, it produces a series of waves that travel outward in all directions on the surface of the water. A wavy line, as shown in Figure 6.1-1, can represent both water waves and electromagnetic energy waves. The top of a wave is called the *crest*, and the bottom of the wave is called the *trough* (pronounced "trof"). The distance from one wave crest to the next wave crest is called the **wavelength**. The height of the crest from the undisturbed surface is the **amplitude** of the wave. The number of waves that pass by a fixed point in a given amount of time is the wave's **frequency**.

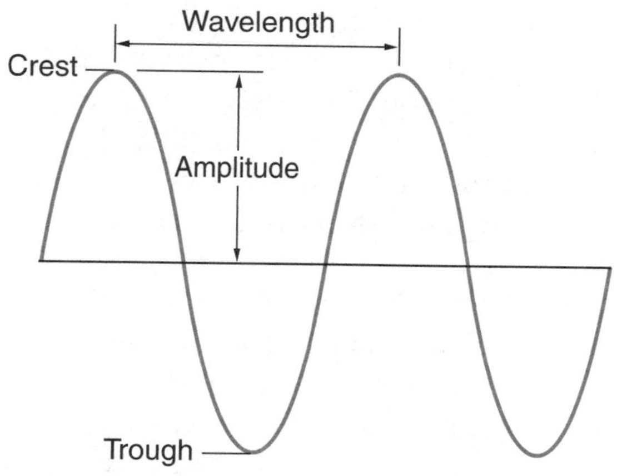

The substance through which waves travel is a **medium**. For instance, water is the medium for the waves produced by the pebble in the example above. Air is the medium through which sound waves travel when your teacher speaks to you. Electromagnetic waves can travel through empty space, so a medium is not necessary. In space, electromagnetic waves travel at a speed of 300,000 km/sec (186,000 mi/sec, "the speed of light"). If you traveled at that speed, you could go around Earth about 7.5 times in one second!

The Electromagnetic Spectrum

The **electromagnetic spectrum** is a continuous band of electromagnetic waves arranged according to wavelength and frequency. (See Figure 6.1-2 on page 170.) Each type of electromagnetic wave differs in wavelength, frequency, and energy. Radio waves have the longest wavelength, the

Figure 6.1-1. Features of a wave.

Figure 6.1-2. The electromagnetic spectrum is a continuous band of electromagnetic waves arranged according to wavelength and frequency.

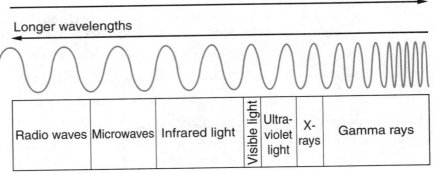

lowest frequency, and the least energy. Gamma rays have the shortest wavelength, the highest frequency, and the most energy.

The characteristics and uses of each type of electromagnetic wave are described below:

- *Radio waves* have the longest wavelength and lowest frequency. The wavelength ranges from longer than a football field to shorter than the length of a football. Radio waves are used to carry AM and FM radio, television, and cell phone signals.
- *Microwaves* have wavelengths about the length of your foot. They are used to heat food, to transmit cell phone signals, and for radar. Doppler radar is used in weather forecasting.
- *Infrared waves* have wavelengths smaller than the head of a pin. Infrared waves are heat waves. We can feel infrared waves from the sun, fire, and radiators. Special infrared lamps are used to heat food and to treat sore muscles. Television remote controls use infrared waves.
- *Light waves* are visible to the human eye. Different wavelengths of light

produce different colors. Red has the longest wavelength and violet has the shortest wavelength. The wavelength of light that is 0.000000510 meter (510 billionths of a meter) produces the color green.

- *Ultraviolet light* waves have shorter wavelengths than those of visible light. Humans cannot see ultraviolet light, but some insects can. Ultraviolet light from the sun causes sunburn that may lead to cancer. Ultraviolet light from distant stars and galaxies helps us to learn about the structure and evolution of these objects.
- *X-rays* have very short wavelengths. They are capable of penetrating skin and muscle. Dentists and doctors use x-rays to examine our teeth and bones. Because large doses of x-rays destroy living cells, they can be harmful to humans.
- *Gamma rays* have the shortest wavelength and highest frequency. Gamma rays have the most energy of any electromagnetic wave. They are used for medical purposes because they kill some types of cancer cells.

Activities

1. There are many types of electromagnetic waves around you. Make a list of five devices in or around your home that use or produce electromagnetic waves. State the function of each device and the type of electromagnetic wave it uses or produces.

2. Infrared light is not visible to the human eye. However, some digital cameras can see infrared light. Press the channel select button on your TV remote control and look into the plastic transmission window. What do you see? Press the channel select button again, but this time view the transmission window using a digital camera. What do you see?

Interesting Facts About Ultraviolet Light

The sun is the principal source of ultraviolet radiation, also called UV light. Like other types of electromagnetic radiation, UV light has a range of frequencies. Most high-frequency UV light is harmful to plant and animal cells. Fortunately, most high-frequency UV light is absorbed in the upper atmosphere, where it makes ozone. (See Lesson 1.3.) Low-frequency UV light reaches Earth's surface and is used in photosynthesis by green plants. Human skin exposed to low-frequency UV light produces vitamin D. Low-frequency UV light also has a tanning effect on the skin.

Excessive exposure to UV light can cause sunburn, early aging of the skin, and skin cancer. In Australia, the slogan "Slip! Slop! Slap!" is used to remind people to protect themselves from the dangers of UV light. The slogan reminds people to *slip* on a shirt, *slop* on sunscreen, and *slap* on a wide-brimmed hat. Apparently, the slogan works. The number of cases of skin cancer has decreased since the use of the slogan. Preventing excessive exposure to the sun while you're young can make for a healthier life when you are older.

Questions

1. Electromagnetic energy travels in waves that can be represented by the example in the diagram below. The top of a wave is called the crest, and the distance from crest to crest is called the wavelength. The diagram shows

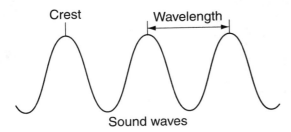

(1) one wave   (3) three waves
(2) five waves   (4) an incomplete wave

2. X-rays are often used for
(1) radio and television broadcasting
(2) communications and cooking food
(3) diagnosing bone injuries
(4) weather radar

3. Overexposure to what type of electromagnetic energy can produce sunburn?
(1) radio waves
(2) light waves
(3) ultraviolet light waves
(4) infrared waves

4. Electromagnetic waves travel in space at
(1) 3 km/sec
(2) 300 km/sec
(3) 1000 km/sec
(4) 300,000 km/sec

5. The electromagnetic spectrum is a continuous band of electromagnetic waves placed in
(1) frequency order
(2) speed order
(3) alphabetical order
(4) temperature order

Thinking and Analyzing

1. Base your answers to the questions on the diagram at right and on your knowledge of science. The diagram represents an electromagnetic wave.
   (a) To what part of the wave is letter A pointing?
   (b) What is the wavelength of this wave?

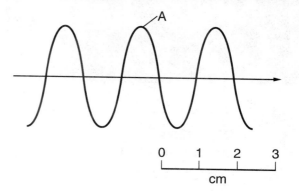

2. Base your answers to the questions on the diagram below and on your knowledge of science. The diagram shows the relationship between wave frequency and wavelength for each type of electromagnetic wave.
   (a) Compare the wavelength and frequency of x-rays and radio waves.
   (b) What is the relationship between wave frequency and wavelength in the electromagnetic spectrum?
   (c) What type of electromagnetic energy from the sun transfers heat to Earth?

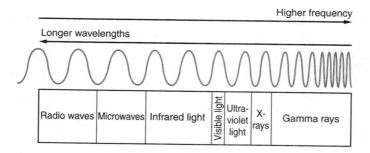

What Is Light?

Objectives
Describe the characteristics of light.

Define the visible spectrum.

Terms
light: a form of radiant energy we can detect with our eyes

visible spectrum: series of colors that make up white light: red, orange, yellow, green, blue, and violet

---

Light Energy

As you learned in the previous lesson, light is a visible form of electromagnetic energy. We can define **light** as a form of radiant energy that we can detect with our eyes. Some properties of light indicate that it consists of tiny bundles of energy. However, other properties indicate that it travels in waves. Light waves move away from a source in all directions. The waves travel in straight-line paths called *rays*. (See Figure 6.2-1.)

Light waves cannot bend around objects. Objects in their path block the light rays and may form shadows.

Like other forms of electromagnetic energy, light can travel through a vacuum. Light from the sun travels through the vacuum of space to reach Earth. At 300,000 km/sec (186,000 mi/sec), sunlight takes about 8 minutes and 20 seconds to reach Earth.

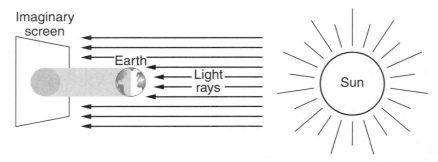

Figure 6.2-1. Light waves travel in straight paths. Shadows are produced when objects block the path of light.

Sources of Light

There are two general sources of light. *Natural* light is produced by nature, and humans produce *artificial* light. The sun is the primary source of natural light on Earth. *Stop and Think:* Is moonlight natural light? *Answer:* Yes. Moonlight is sunlight reflected from the surface of the moon. Lightning is another source of natural light. We also receive a very small amount of natural light from distant stars.

One method of producing artificial light is by heating an object until it glows. A hot, glowing object is said to be *incandescent* (in-kan-DES-nt). Adding heat to an iron nail will cause it to produce light. The incandescent nail will become hotter and hotter, from red-hot, to white-hot, and eventually to blue-hot. The heated filament in an electric lightbulb and the burning wick of a candle are examples of incandescent objects.

Another method of producing artificial light is by passing an electric current through a gas that is enclosed in a glass tube. This is how neon signs are made. If neon gas is used in the tube, a red light is produced. However, neon light signs can use other gases, such as helium, carbon dioxide, and mercury. Each

gas produces its own unique color. At night, many colorful neon signs can be seen on storefronts and on theater marquees. *Stop and Think:* Are signs that contain other gases still called neon lights? *Answer:* Yes. Even though they do not contain neon gas, they are still called neon lights.

Color

The visible light portion of the electromagnetic spectrum consists of waves that can be seen by the human eye. Visible light falls within a very narrow range on the electromagnetic spectrum called the **visible spectrum**. White light that makes up the visible spectrum can be separated into colors. These colors are red, orange, yellow, green, blue, and violet (ROYGBV). (See Figure 6.2.2.) *Stop and Think:* Have you ever seen the visible spectrum? *Answer:* You may have seen the visible spectrum in a rainbow, in light reflected by a CD, or when bright light is passed through a glass prism.

Each color within the visible spectrum differs in wavelength and frequency. Red light has the longest wavelength and lowest frequency, and violet light has the shortest wavelength and highest frequency.

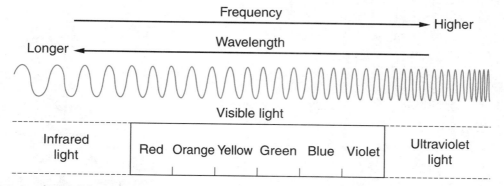

Figure 6.2-2. The visible spectrum is a portion of the electromagnetic spectrum. It consists of the colors red, orange, yellow, green, blue, and violet.

The color that you see when you look at an object is the result of some wavelengths bouncing off the object and other wavelengths being absorbed by the object. The wavelengths that bounce off an object determine the color of the object. You may wonder why a car looks blue. When white light from the sun strikes a blue car, all the wavelengths of the visible spectrum are absorbed except the blue wavelength. The blue wavelength bounces off the car and travels to your eyes. Therefore, you see a blue car.

SKILL EXERCISE—*Reading Comprehension and Sky Colors*

*B*ase your answers to the questions on the following passage and on your knowledge of science.

Sky Colors

You may wonder why the sky usually appears blue during the day and sometimes red or orange at sunrise or sunset. Incoming sunlight is white. White light consists of the visible spectrum: red, orange, yellow, green, blue, and violet. Red is at the long-wavelength end of the visible spectrum, while blue and violet are at the short-wavelength end.

Air molecules in the atmosphere scatter sunlight. However, all wavelengths of sunlight are not scattered the same. Air molecules scatter short wavelengths more than long wavelengths. Therefore, blue and violet, which have a short wavelength, are scattered more efficiently. Our eyes are more sensitive to blue light than to violet light. So the sky appears blue rather than violet.

Beautiful sunrises and sunsets often appear in the sky. The larger particles in the atmosphere such as dust, smoke, and cloud droplets, and the low angle of the sun at sunrise and sunset scatter all wavelengths of light. However, the shorter wavelengths (blue and violet) are scattered away from you, and the longer wavelengths (red, orange, and yellow) are scattered toward you. This often produces a colorful sky at sunrise and sunset.

Questions

1. What colors in the visible spectrum have the shortest wavelengths?

2. What causes the sky to appear blue?

3. Volcanic ash is a fine dust that floats in the atmosphere for long periods of time after a volcanic eruption. How does volcanic ash in the atmosphere affect the color of the sky?

Activities

1. A shadow is produced when an object blocks light. There are two parts to a shadow. The darker central portion of a shadow is called the *umbra* (UM-bruh), and the lighter outer portion is called the *penumbra* (puh-NUM-bruh). Place a bright lamp in a dark room. Hold a ball between the wall and the lamp and observe the shadow that is produced. Draw a labeled diagram in your notebook showing how you produced the shadow. Label the umbra and penumbra. Describe what happens to the penumbra when you move the ball closer to the wall.

2. The color of a neon sign depends on the type of gas used in its glass tubes. More than 150 different colors can be produced. The Internet site *http://inventors.about.com/library/weekly/aa980107.htm* presents the history of neon signs; describes how they are made; shows the different colors; and has many pictures of different types of neon signs. See how many different neon signs and colors you can locate in your neighborhood shopping area at night.

Interesting Facts About Light From Stars

Stars are very bright objects. However, they appear dim because they are so far away. Starlight travels at 300,000 km/sec (186,000 mi/sec) across space. Even at that great speed, it takes a long time for starlight to reach Earth.

Consider the fact that when we look at a star we are looking back in time. For example, Polaris, the North Star, is a typical star in our nighttime sky. It is about 4085 trillion km (2538 trillion miles) away from Earth. It takes starlight from Polaris about 430 years to reach Earth. That means the light that left Polaris in 1578 reaches Earth in 2008. If Polaris exploded today, the people who live on Earth 430 years from now would see the explosion. Or it could have already exploded and we don't know about it yet!

Questions

1. The visible spectrum consists of the
   (1) shortest electromagnetic wavelengths
   (2) longest electromagnetic wavelengths
   (3) colors red, orange, yellow, green, blue, and violet
   (4) colors white, red, yellow, blue, and black

2. When white light strikes a green wall, all the visible wavelengths
   (1) are absorbed
   (2) are absorbed except the green wavelength, which bounces off the wall
   (3) bounce off the wall
   (4) bounce off the wall except the green wavelength, which is absorbed

3. What property of red light is different from violet light?
   (1) speed   (3) temperature
   (2) frequency   (4) pressure

4. What is the primary source of light on Earth?
   (1) sunlight
   (2) starlight
   (3) light from fire
   (4) incandescent light

5. Shadows are formed when
   (1) light changes speed
   (2) light changes direction
   (3) an object blocks light
   (4) an object reflects light

Thinking and Analyzing

Base your answers to the questions on the diagram below and on your knowledge of science. The diagram shows the frequency and wavelength of the visible spectrum.

1. Which color in the visible spectrum has the longest wavelength?

2. What is the relationship between frequency and wavelength?

3. Compare the wavelength and frequency of yellow light and blue light.

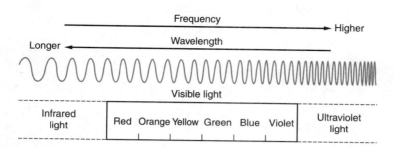

What Are the Properties of Light?

Objectives
Describe what can happen when light strikes a surface.

Explain what causes the refraction (bending) of light.

Terms
reflection: the bouncing of light rays off a surface

absorption: the process by which light rays are absorbed and transformed into heat

transmission: the passing of light rays through an object

refraction: the bending of light rays when they pass from one medium to another

lens: a piece of transparent glass or plastic that has a curved surface

---

Light Can Be Reflected, Absorbed, or Transmitted

When light strikes the surface of an object, three things can happen to the light rays. (See Figure 6.3-1.)

1. Light rays can bounce off the surface of the object. This is called **reflection**. A shiny, metal surface reflects most of the light that strikes it.
2. Light rays can be **absorbed** by the object and transformed into heat energy. Much of the light that strikes a blacktop road is absorbed and transformed into heat energy.
3. Light rays can pass through the object. This is called **transmission**. Clear glass allows most light that strikes it to pass through it.

We see objects because they produce light or reflect light. Light rays enter our eyes, and our brain interprets what we see. Accurate reflections are produced by smooth

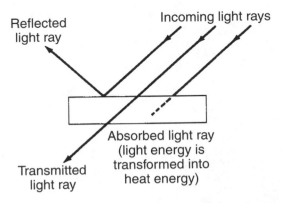

Figure 6.3-1. Depending on the type of surface it strikes, light may be reflected, absorbed, or transmitted.

and shiny surfaces. A mirror gives an accurate reflection because it has a smooth, shiny surface. A wall in your home produces a poor reflection because its rough surface scatters light in all directions.

Colored objects absorb light to varying degrees. Dark-colored objects absorb more sunlight than do light-colored objects. Because absorbed light changes to heat energy, dark-colored objects heat up more than light-colored objects when exposed to sunlight. For this reason, people usually wear light-colored clothing to keep cool during hot, sunny weather. **Stop and Think:** Why should you wear dark-colored clothes on a cold, sunny day? **Answer:** Dark-colored clothes absorb heat better from sunlight than do light-colored clothes.

Materials also differ in their ability to transmit light. *Transparent* materials, such as window glass, permit almost all of the incoming light to pass directly through them. *Translucent* (tranz-LOO-sent) materials, such as wax paper, let some light pass through, but scatter light rays, so images are not transmitted clearly. *Opaque* (oh-PAKE) objects, like wood and iron, do not allow any light to pass through them.

Refraction of Light

Refraction is the bending of light rays caused by light passing from one medium into another. For example, when light from an object passes from water into air, or from air into glass, the light is refracted or bent, causing the object to look larger, smaller, or displaced. Refraction causes a pencil in a glass of water to look broken or bent where it enters the water. (See Figure 6.3-2.) Light rays reflecting from the pencil are refracted

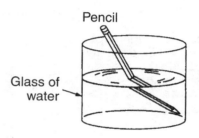

Figure 6.3-2. The bending or refraction of light rays as they pass from one medium to another makes the pencil look bent or broken.

as they pass out of the glass from the water into the air. The principles of refraction are used in making lenses for cameras, eyeglasses, and telescopes.

If a beam of white light is passed through a glass prism, the beam is refracted and dispersed (spread out and separated) into the visible spectrum. (See Figure 6.3-3.) The separation occurs because each color is refracted differently due to its different wavelength. Red has the longest wavelength and is refracted the least. Violet has the shortest wavelength and is refracted the most.

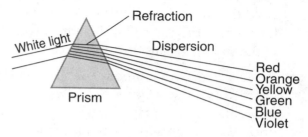

Figure 6.3-3. White light passing through a prism is refracted and dispersed, producing the visible spectrum.

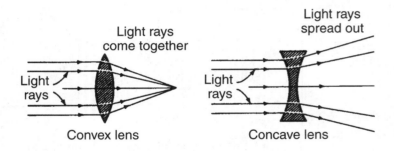

Figure 6.3-4. The shape of a lens determines how it bends rays of light.

Lenses

A **lens** is a piece of transparent glass or plastic that has curved surfaces. The curved surfaces refract light rays that pass through the lens. The shape of a lens determines how it bends light, as shown in Figure 6.3-4. A lens with surfaces that curve outward (*convex*) bends light rays inward so that they focus at a common point. A lens with surfaces that curve inward (*concave*) causes light rays to spread out.

Images of objects seen through lenses may be larger than, the same size as, or smaller than the objects themselves. For instance, the lens of a camera forms smaller images of objects. A photocopy machine has a lens that forms images the same size as the original object. Binoculars and telescopes contain lenses that magnify objects, making them appear larger. *Stop and Think:* What does the lens in a microscope do to an image? *Answer:* The lens in a microscope makes the image appear larger.

Activities

1. Have you ever stood in front of a fun-house mirror? These mirrors are designed to make you look funny by changing the shape of your head and body. You can get the same effect using a shiny silver spoon. The larger the spoon, the better it works. Look at your reflection in the spoon. Hold the spoon horizontally and look at your image on the concave surface of the spoon. Turn the spoon over and look at your image on the convex surface. Hold the spoon vertically and repeat the procedure. Take notes and draw four simple diagrams that show how you look in each of the four positions of the spoon. What other objects can you use to do this activity?

2. Produce your own rainbow. You need a bright sunny day and a garden hose attached to a water faucet. Stand with your back to the sun and use the hose to produce a fine spray of water in the direction of your shadow. Spray in an upward direction and let the mist fall to see a larger rainbow. Note the order of the colors in your rainbow. Is red located on the top or on the bottom of your rainbow? Draw a labeled diagram in your notebook to show how you produced your rainbow.

Interesting Facts About Rainbows

Have you ever wondered how a rainbow forms? Some stories say there is a pot of gold at the end of a rainbow. Unfortunately, a rainbow is an optical illusion that disappears as you approach it. Sorry to say, you will never find the pot of gold.

Rainbows form under specific weather conditions. The sun must be low in the sky and it has to be raining in the opposite direction away from the sun. You must have your back toward the sun and be facing the rain. The falling raindrops refract and reflect the sun's rays back to you. (See Figure 6.3-5.)

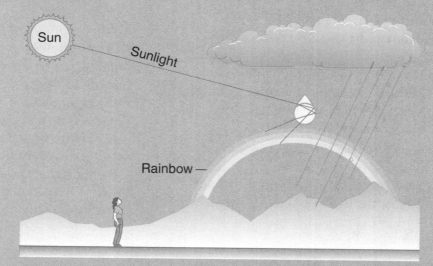

Figure 6.3-5. Rainbows form under specific weather conditions.

A ray of white sunlight enters a raindrop and is divided into different colors by refraction and dispersion. This is the same process that causes white light to split into individual colors when it passes through a glass prism. Within the raindrop, the refracted light is reflected from the back wall of the raindrop. The reflected light passes back into the air and back toward you. From Earth, we see a continuous spectrum of colors forming an arc in the sky.

Questions

1. We can see most objects because they
   (1) bend light   (3) absorb light
   (2) reflect light   (4) transmit light

2. A magnifying glass is used to make objects appear larger. What process does a magnifying glass use to make objects appear larger?
   (1) reflection   (3) absorption
   (2) transmission   (4) refraction

3. A lens can be used to produce images that are
   (1) larger than the original object
   (2) the same size as the original object
   (3) smaller than the original object
   (4) all of the above

4. Skiers often wear sunglasses while they are skiing because snow
   (1) radiates light   (3) conducts light
   (2) absorbs light   (4) reflects light

5. When using the apparatus shown in the diagram below, the student could see the flame only if all three holes were lined up.

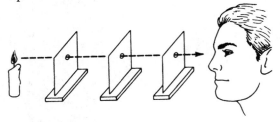

What property of light does this demonstrate?
(1) Light rays are reflected from smooth, shiny surfaces.
(2) Light rays are absorbed as heat by dark-colored surfaces.
(3) Light rays travel in straight paths.
(4) Transparent objects transmit most of the light that strikes them.

Thinking and Analyzing

1. Base your answers to the questions on the diagram at the right and on your knowledge of science. The diagram shows how a glass prism affects white light.
(a) What happens to the white light as it passes through the prism?
(b) Which color of the visible spectrum is bent the most?
(c) Explain why white light is bent when it strikes the prism.

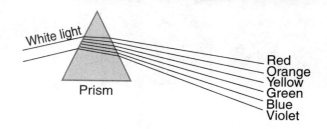

2. Base your answers to the questions on the diagrams below and on your knowledge of science. The diagrams show what happens to a light ray when it strikes a surface.

(a) Which surface represents a mirror?
(b) Describe what is happening to the light ray in diagram C.
(c) What type of energy transformation is occurring in diagram B?

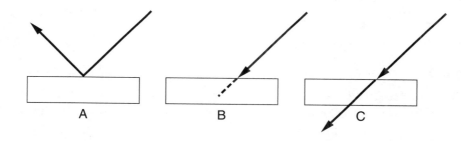

6.4 What Is Sound?

Objectives

Describe the characteristics of sound.

Compare sound waves and electromagnetic waves.

Explain how air temperature affects the speed of sound.

Terms

sound: a form of mechanical energy produced by a vibrating object

transverse wave: a wave that vibrates at a right angle (up and down) to the direction in which the wave is traveling

longitudinal (lahn-juh-TOO-duhn-uhl) **wave:** a wave that vibrates back and forth (push and pull) in the same direction as its direction of travel

Sound Energy

Sound is a form of mechanical energy produced by a vibrating object. A sound wave is created when an object vibrates or moves rapidly back and forth. This back and forth motion pushes and pulls the surrounding matter, producing alternating layers of molecules that are closer together (compressed) and molecules that are farther apart (expanded). *Stop and Think:* Why is sound considered a form of mechanical energy? *Answer:* Mechanical energy is the energy of a moving object, and sound waves are moving molecules. Sound waves spread out in all directions from their source, somewhat like the circular ripples that are produced when you toss a pebble into a pool of calm water.

Objects that produce sound include bells, radio speakers, guitar strings, and anything else that can vibrate. (See Figure 6.4-1.) For instance, the sound of your voice is caused by the vibrating vocal cords in your throat. If you place your hand on your throat while you speak, you can feel the vibrations that produce the sound. (Go ahead. Try it!)

To be heard, a sound must be transmitted from a source to your ears. The substance that sound travels through is called a *medium*. A medium can be a solid, a liquid, or a gas. Most sounds we hear travel through air, a gas. Sound waves cannot travel through a *vacuum* because there are no molecules of matter in a vacuum to transmit the sound. *Stop and Think:* If an astronaut banged a pot on the

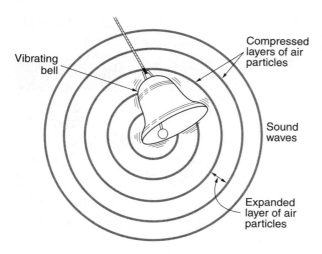

Figure 6.4-1. A vibrating bell produces a series of sound waves that travel through air.

moon, could he hear it? **Answer:** He could not hear the sound because the moon has no atmosphere, and sound waves need a medium to travel through.

Types of Waves

A wave that vibrates at a right angle (up and down) to the direction in which the wave is traveling is a **transverse wave**. (See Figure 6.4-2.) Electromagnetic waves and ocean waves are transverse waves. Transverse waves can be demonstrated by tying one end of a rope to the back of a chair and shaking the untied end of the rope up and down. Although you see the wave moving along the rope, the actual material of the rope does not move forward, but rather moves up and down with each passing crest and trough.

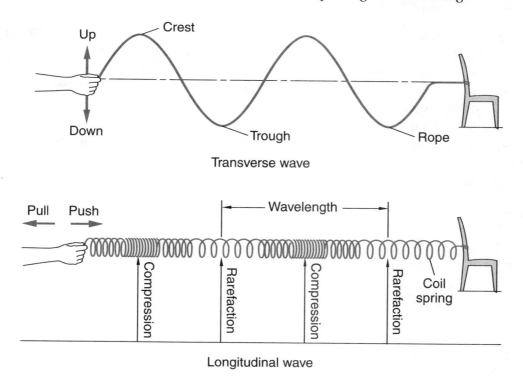

Figure 6.4-2. A rope can be used to demonstrate a transverse wave (top). A coiled spring can be used to demonstrate a longitudinal wave (bottom).

On the other hand, a wave that vibrates back and forth (push and pull) in the same direction as its direction of travel is a **longitudinal wave**. Sound waves are longitudinal waves. A longitudinal wave can be demonstrated by using a long coiled spring, like a Slinky™ toy. Attach the spring to the back of a chair and stretch the spring out, then push it in. You see a series of waves pass through the spring. The area where the spring coils push close together is called *compression*. The area where the coils pull apart, or spread out, is called *rarefaction* (rare-ah-FAK-shun). The wavelength is measured from compression to compression. Figure 6.4-2 illustrates the two types of waves.

The Speed of Sound

The speed of sound depends primarily on the density of the substance, or medium, through which it is passing. The denser the medium, the faster sound waves travel through it. Generally, sound travels fastest through solids, which have the greatest density, and slowest through gases, which have the least density. Table 6.4-1 gives the speed of sound through several substances.

Table 6.4-1. Speed of Sound Through Different Substances (at 25°C)

Medium	State	Speed (m/sec)
Iron	Solid	5200
Glass	Solid	4540
Water	Liquid	1497
Air	Gas	346

To a lesser extent, temperature also affects the speed of sound. In air, the higher the temperature, the faster sound waves travel through it. (See the Skill Exercise on Reading a Graph on page 189.)

Although the speed of sound can vary, it is always much slower than the speed of light. During a thunderstorm, for instance, a lightning bolt produces a flash of light and a clap of thunder at the same time. *Stop and Think:* Assuming you are several kilometers away, which will you sense first, the lightning or the thunder? *Answer:* The speed of light is so fast that the light reaches us almost instantly. The sound of thunder travels much more slowly, so we usually hear the thunder several seconds after we see the lightning.

SKILL EXERCISE—*Reading a Graph*

ound travels at different speeds through different substances. The speed of sound is faster in solids, such as stone or metal, and slower in liquids and gases, such as water and air. The speed of sound through air is also affected by air temperature. The graph to the right shows the relationship between air temperature and the speed of sound. Study the graph and then answer the following questions.

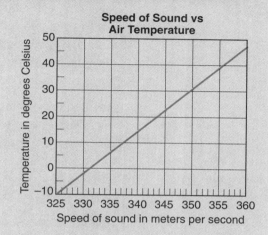

Questions

1. At which temperature does sound travel at a speed of 340 meters per second?
(1) 6°C   (2) 10°C   (3) 15°C   (4) 22°C

2. At a temperature of 31°C, sound travels about
(1) 345 meters per second
(2) 350 meters per second
(3) 355 meters per second
(4) 340 meters per second

3. What does the graph suggest about the relationship between air temperature and the speed of sound?
(1) As air temperature decreases, the speed of sound increases.
(2) As air temperature decreases, the speed of sound remains the same.
(3) As air temperature increases, the speed of sound increases.
(4) As air temperature increases, the speed of sound remains the same.

4. The speed of sound is
(1) the same in a solid, liquid, and gas
(2) fastest in a solid
(3) fastest in a liquid
(4) fastest in a gas

Interesting Facts About Radar, Sonar, and Bats

Radar stands for **Ra**dio **D**etection **a**nd **R**anging. Radar is used to detect and determine the direction of objects. Radio waves are sent out by a transmitter, reflected back by the target, and captured by a receiver. If you know the speed of the waves and the time between transmission and reception, you can determine the distance to the object. Many professions use radar. Radar is used by airplane pilots to detect other planes, by police officers to identify speeding cars, and by your local weather forecaster (meteorologist) to locate and predict rain and other forms of precipitation.

Sonar stands for **So**und **Na**vigation and **R**anging. Sonar is used to detect underwater objects. It is similar to radar, but uses sound waves instead of radio waves. *Echo sounding* is a term used to describe the use of sonar to measure the depth of the ocean. Sonar was used to discover the features of the ocean floor that contributed to the theory of plate tectonics.

About 800 species of bats use sound to navigate and to locate prey. Bats have a built-in sonar system referred to as *echolocation* (eh-koh-lo-KAY-shun). Bats send out sonar impulses that are reflected from surrounding objects. A returning echo from a flying insect provides the bat with the speed, direction, and size of its prey. After repeating this process many times, the bat eventually swoops down, captures, and eats its prey.

Activities

Sound travels through gases, solids, and liquids. The first two activities demonstrate sound traveling in a gas and a solid. The third activity asks you to design a demonstration to show sound can travel in a liquid.

1. Gas: Securely place plastic wrap over the open end of a coffee can. Sprinkle some salt on top of the plastic wrap. Hold a large pot near the side of the coffee can and tap the pot with a wooden spoon. What happens to the salt on top of the plastic wrap? Continue to tap the pot as you slowly move it away from the coffee can. How far away can you move from the coffee can and still get a reaction? Place the coffee can on the floor and hold the pot above it. Tap the pot while slowly moving it higher and higher above the coffee can. Is the reaction better with the pot on the side of the coffee can or above the coffee can?

2. Solid: Place your ear against a wall, cover your other ear, and gently tap the wall about 15 cm (6 inches) in front of your face. Can you hear the sound? Repeat this procedure using a wooden table, a glass window, and a large paperback book. Describe the sound you hear in each surface. How are they different?

3. Liquid: Design a demonstration that shows sound can travel through a liquid.

Questions

1. Sound is produced by
   (1) expansions   (3) reflections
   (2) contractions   (4) vibrations

2. When using the apparatus in the diagram below, the student could not hear the ringing bell after the air was pumped out of the bell jar.

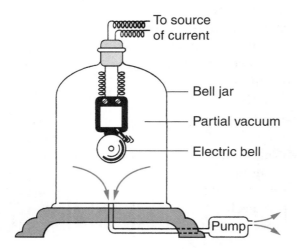

To source of current

Bell jar

Partial vacuum

Electric bell

Pump

This demonstrates that sound
(1) can travel through the glass bell jar
(2) cannot travel through the glass bell jar
(3) cannot travel through a vacuum
(4) can travel through a vacuum

3. At which air temperature would sound waves travel the fastest?
   (1) 0°C   (3) 10°C
   (2) 5°C   (4) 20°C

4. Compared to the speed of light, the speed of sound is
   (1) much faster
   (2) much slower
   (3) nearly the same
   (4) a little bit slower

5. Sound is a form of
   (1) radiant energy
   (2) mechanical energy
   (3) chemical energy
   (4) heat energy

Thinking and Analyzing

1. Base your answers to the questions on the passage, the figure, and your knowledge of science. An echo is produced when sound waves reflect from a surface and return to the sender.

   Sound travels about 340 m/sec (1115 feet/second) in air. Josh yelled "hello" across a canyon. The sound traveled across the canyon and returned as an echo. Josh heard the echo two seconds after he yelled "hello."

   (a) How far did the sound travel?
   (b) What is the distance across the canyon?
   (c) If the canyon were 1020 m (3345 feet) across, how long would it take for Josh to hear the echo?

2. Sound and light are two forms of energy. Sound is a form of mechanical energy and light is a form of radiant energy.
   (a) State two ways that sound and light are similar.
   (b) State two ways that sound and light are different.

6.5 What Are the Properties of Sound?

Objectives
Identify the properties of a sound wave.

Explain how motion affects the frequency of sound.

Terms
hertz (HURTS): unit used to measure the frequency of waves

pitch: the level of frequency of a sound, that is, how high or low it sounds, which is determined by the frequency of the sound wave

intensity: the amount of energy in a sound wave, that is, how loud it sounds, which is determined by the amplitude of the sound wave

decibel: unit used to measure the intensity or loudness of sound

Doppler (DAH-pluhr) **effect:** a change in wave frequency caused by the relative distance between the sound source and the listener

Frequency and Pitch

Each sound has it own unique properties. A nearby car honks a horn. A train blows a whistle. Someone loudly calls your name. In each case, we recognize the sound because of its properties. In this lesson, we will discuss two physical properties of sound waves—*pitch* and *intensity.*

In Lesson 6.1, we learned that frequency is the number of waves that pass a fixed point in a given amount of time. Wave frequency is commonly expressed as the number of waves per second. The frequency of a sound is measured in units called **hertz** (Hz). One hertz equals 1 wave per second. A frequency of 120 Hz means 120 waves pass a fixed point in 1 second.

Humans can hear sounds that range from 20 to 20,000 Hz. A normal speaking voice has a range between 100 to 1000 Hz and a wavelength of 0.3 to 3.5 meters.

Pitch describes how high or low a tone sounds. The pitch of a sound wave is determined by its frequency. A high-frequency sound has a high pitch, and a low-frequency sound has a low pitch. (See Figure 6.5-1a.) A violinist or guitar player changes pitch by sliding his or her fingers up and down the instrument strings. This causes the length of the strings on the instrument to change. A long string vibrates less rapidly, has a lower frequency, and produces a low pitch. A short string vibrates more rapidly, has a higher frequency, and produces a higher pitch.

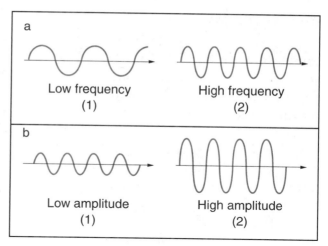

Figure 6.5-1. (a) Wave a(2) has a higher frequency and higher pitch than wave a(1). (b) Wave b(2) has a higher amplitude and a higher intensity than wave b(1).

Amplitude and Intensity

Intensity is the amount of energy in a sound wave. In Lesson 6.1, we learned that the height of the wave crest is called the amplitude. A wave's amplitude is a measure of the energy it contains. When comparing two sound waves, the wave with the higher amplitude has more energy and is louder than the wave with the lower amplitude. (See Figure 6.5-1b.)

Loud sound waves carry a greater distance because they contain more energy. A shout carries much farther than a whisper. Call to friends across the street. First whisper and see if they can hear you. Try calling them with a normal talking voice. You will most likely need to shout for them to hear you. Shouting produces sound waves that have more energy. More energy is needed for sound to travel greater distances.

The intensity or loudness of sound is measured in units called **decibels** (dB). The softest sound most humans can hear has an intensity of about 20 decibels. The intensity of normal conversation is about 65 decibels. Sound intensities greater than 120 decibels

can be harmful to human ears. Table 6.5-1 lists the approximate intensity of different sounds in decibels.

Table 6.5-1. Intensity of Sounds

Sound	Intensity Level (dB)
Threshold of hearing (youth)	0
Rustling leaves, buzzing mosquito	10
Whisper, quiet room in house	20
Soft music	30
City background noise	45
Normal conversation	60
Busy street traffic, laughter	70
Vacuum cleaner	80
Subway train, motorcycle	90
Loud music	100
Rock concert, power tools	110
Threshold of pain	130
Jet takeoff	140
Eardrum damage	160

Doppler Effect

To a stationary observer, the frequency of sound waves produced by the siren on an ambulance increases or decreases, depending on whether the ambulance is moving toward or away from the listener. This change in sound frequency, called the **Doppler effect**, is recognized as a change in pitch. When a blaring ambulance siren moves toward you, the sound wavelengths shorten and crowd together. This produces a higher frequency and a higher pitch. When the siren moves away from you, sound wavelengths lengthen or spread out. This produces a lower frequency and a lower pitch. (See Figure 6.5-2.)

The Doppler effect can also be used with electromagnetic waves to detect and measure the motion of objects. Police use the Doppler effect of radio waves (radar) to detect the speed of moving cars. Doppler radar is used in weather forecasting to track the movement of precipitation. In medicine, doctors use the Doppler effect of ultrasonic waves to monitor the circulation of your blood. The motion of stars can be detected and measured using the Doppler effect on light waves.

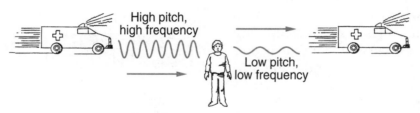

Figure 6.5-2. When the siren of the ambulance moves closer to you, the frequency and the pitch increase. When the siren of the ambulance moves away from you, the frequency and the pitch decrease.

SKILL EXERCISE—*Reading a Range Graph*

Humans hear sound frequencies in the range of 20–20,000 hertz, or 20–20,000 vibrations per second. Animals also hear sound in a frequency range specific to the type of animal. Some animals can hear sounds below 20 hertz. These sounds are called *infrasonic*. Some animals can hear sounds above 20,000 hertz. These sounds are called *ultrasonic*. The graph below shows the approximate hearing frequency range of various animals.

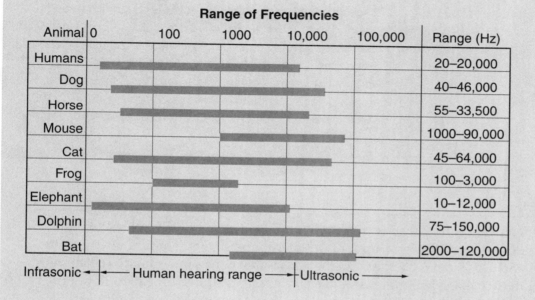

Range of Frequencies

Animal	0	100	1000	10,000	100,000	Range (Hz)
Humans						20–20,000
Dog						40–46,000
Horse						55–33,500
Mouse						1000–90,000
Cat						45–64,000
Frog						100–3,000
Elephant						10–12,000
Dolphin						75–150,000
Bat						2000–120,000

Infrasonic ◄─┼─ Human hearing range ─►┤Ultrasonic ───►

Questions

1. What animal can hear infrasonic sound?

2. Dolphins generally produce sound in the 18,000–20,000 Hz range. What animals in the graph cannot hear dolphins?

3. A dog whistle can be heard by a dog but not by humans. A dog whistle produces sound waves within what frequency range?

Interesting Facts About the Loudest Sounds on Earth

What is the loudest sound ever made on Earth? One of the loudest natural sounds ever to occur on Earth was caused by a meteorite impact 65 million years ago. Scientists believe that the impact resulted in the extinction of the dinosaurs.

Volcanic eruptions are the loudest sounds that Earth produces. The eruption of the Krakatoa volcano in 1883 could be heard as far away as 8000 km (5000 miles). Major volcanic eruptions produce sounds greater than 300 decibels. The eruption of Mount Mazama (Crater Lake, Oregon) about 7000 years ago and Mount Tambora (Indonesia) in 1815 were loud enough to be heard by people great distances away.

Atomic and nuclear bombs create the loudest sounds produced by humans. The sound from these explosions approaches 300 decibels. The sound of a blue whale travels hundreds of kilometers underwater and is considered the loudest sound made by any animal on Earth. The sound of military rifles and jet engines can produce sounds as high as 150 decibels. Recent studies show that subway noise levels can reach 106 decibels. Extended exposure to sound at this level can impair or damage your hearing.

Activities

1. Pour different amounts of water in eight identical glasses or glass bottles. Place the glasses in order, from the least amount of water to the greatest amount of water. Use a spoon and gently tap each glass. Do the glasses make the same sound? The amount of water in each glass determines its pitch. Try to play a simple tune like "Row, Row, Row Your Boat," "Twinkle, Twinkle, Little Star," or "Mary Had a Little Lamb."

2. Try to estimate the intensity of different sounds. Although location and conditions vary greatly, give it a try. Record your location and use Table 6.5-1 on page 195 to help determine the approximate intensity in decibels of the following:

(a) your school cafeteria at lunchtime

(b) clapping your hands

(c) hitting a pot with a metal spoon

(d) the music on your iPod or radio

(e) the commercials on TV

(f) the sound in the movies

(g) riding a subway train

(h) cheering fans at a sporting event

Questions

1. The frequency of a sound wave determines the sound's
   (1) loudness   (3) distance
   (2) pitch   (4) motion

2. The amplitude of a sound wave determines the sound's
   (1) intensity   (3) distance
   (2) pitch   (4) motion

3. What unit is used to measure the frequency of a wave?
   (1) centimeters/second
   (2) decibels
   (3) meters/second
   (4) hertz

4. Intensity is to decibel as pitch is to
   (1) amplitude   (3) hertz
   (2) Doppler effect   (4) sound

Thinking and Analyzing

1. Base your answers to the questions on the passage and on your knowledge of science.

   A passing train blows a whistle at a constant frequency as it approaches, crosses, and leaves a road crossing. A person standing at the road crossing notices that the pitch of the whistle changes.

   (a) Describe what happens to the frequency and pitch of the sound waves as the train approaches the person.
   (b) Describe what happens to the frequency and pitch of the sound waves as the train moves away from the person.
   (c) What is this change in sound frequency called?

2. Base your answers to the questions on the passage below and on your knowledge of science.

   You can determine the distance from your location to a thunderstorm with a simple mathematical calculation. From the safety of your home, watch for a flash of lightning, then count the number of seconds that pass until you hear thunder. You simply divide this number by 3 to determine your distance from the storm in kilometers. (Actually, you determine the distance to the lightning.)

   (a) How far away is the thunderstorm if the time difference between the flash of lightning and the thunder is 6 seconds?
   (b) How many seconds would pass between the flash of lightning and the thunder if the storm were 5 kilometers away?

Review Questions

Term Identification
Each question below shows two terms from Chapter 6. One of the terms is defined.
(1) Choose the term that matches the definition.
(2) Describe how the two terms are different. Following each term is the section (in parenthesis) where the description or definition of that term is found.

1. *Wavelength (6.1) — Amplitude (6.1)*
   The distance from one wave crest to the next wave crest

2. *Electromagnetic spectrum (6.1) — Visible spectrum (6.2)*
   A continuous band of all electromagnetic waves

3. *Radio wave (6.1) — Gamma ray (6.1)*
   Electromagnetic wave with a short wavelength and high frequency

4. *Reflection (6.3) — Refraction (6.3)*
   The bending of light rays that pass from one medium to another

5. *Absorption (6.3) —Transmission (6.3)*
   The passing of light waves through an object

6. *Longitudinal wave (6.4) —Transverse wave (6.4)*
   A wave that vibrates at a right angle (up and down) to the direction in which the wave is traveling

7. *Pitch (6.5) — Intensity (6.5)*
   Determined by the frequency of a wave

8. *Decibel (6.5) — Hertz (6.5)*
   Unit used to measure the frequency of waves

Multiple Choice (Part 1)
Choose the response that best completes the sentence or answers the question.

1. Which statement describes how sound waves and light waves are similar?
   (1) Both waves can bend around objects.
   (2) Both waves originate from an energy source.
   (3) Both waves travel at the same speed.
   (4) Both waves are a series of compressions and rarefactions.

2. A blast of dynamite set off by a roadwork crew produced a bright flash of light and a loud explosion. A person standing two kilometers away with a clear view of the work site
   (1) hears the sound first, then sees the flash
   (2) sees the flash first, then hears the sound
   (3) sees the flash and hears the sound at the same time
   (4) hears the sound but does not see the flash

3. What happens to light rays that pass through two transparent substances of different densities, such as air and a magnifying glass?
(1) The light rays are refracted.
(2) The light rays are reflected.
(3) The light rays are absorbed.
(4) The light rays travel in a straight line.

4. While visiting an historic fort during his vacation, John watched a demonstration involving the firing of a cannon. When the cannon fired, the sound traveled to John and he heard the sound. What is the medium through which the sound waves traveled?
(1) the cannon   (3) John's ears
(2) the air   (4) the cannonball

5. A common characteristic of sound waves is that they
(1) are created by vibrations
(2) travel in straight lines toward the source
(3) travel fastest through empty space
(4) move at the speed of light

6. Light that strikes an opaque object is reflected and
(1) absorbed
(2) transmitted
(3) refracted
(4) compressed

7. A microscope lens makes objects look larger because the lens
(1) reflects light
(2) absorbs light
(3) transmits light
(4) refracts light

8. When you view a rainbow, the sun is located
(1) to your right
(2) to your left
(3) behind you
(4) in front of you

9. An example of an artificial incandescent light source is
(1) a neon light
(2) a lightbulb with a glowing filament
(3) the sun
(4) the moon

10. A glass prism can be used to demonstrate the
(1) different types of electromagnetic energy in white light
(2) reflection of light
(3) refraction and dispersion of white light into the visible spectrum
(4) speed of light

11. Light is the
(1) visual portion of the electromagnetic spectrum
(2) portion of the electromagnetic spectrum with the longest wavelengths
(3) portion of the electromagnetic spectrum with the shortest wavelengths
(4 largest portion of the electromagnetic spectrum

12. Sound is a form of
(1) heat energy
(2) electromagnetic energy
(3) chemical energy
(4) mechanical energy

Thinking and Analyzing (Part 2)

1. Base your answers to the questions on the diagram below and on your knowledge of science. The diagram shows a person observing a fish located at position B below the surface of the water. The person sees the fish at position A.

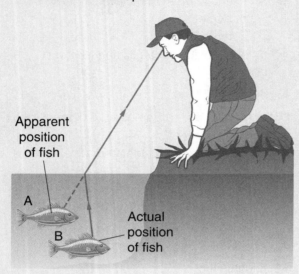

Apparent position of fish

A

B

Actual position of fish

(a) Explain why the apparent position of the fish is different from the actual position of the fish.
(b) What two mediums do the light waves pass through to reach the person's eyes?

2. Base your answers to the questions on the diagram at the bottom of the page and on your knowledge of science. The diagram represents the different forms of electromagnetic energy.

(a) What feature best distinguishes one form of electromagnetic energy from another?
(b) Which color in the visible light portion of the electromagnetic spectrum has the longest wavelength?
(c) Which type of electromagnetic energy contains the most energy?

Electromagnetic Spectrum

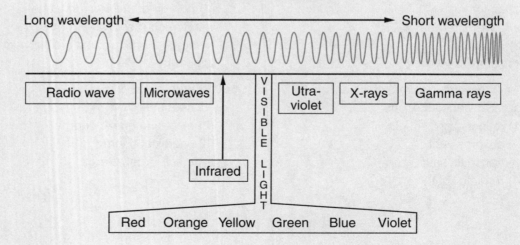

3. Light and sound are two forms of energy. Compare the following characteristics of light and sound energy:
   (a) the speed of the waves
   (b) the medium the waves can travel through
   (c) type of waves
4. Base your answers to the questions on the diagram below and on your knowledge of science. The woman observes white sunlight striking the red stop sign.

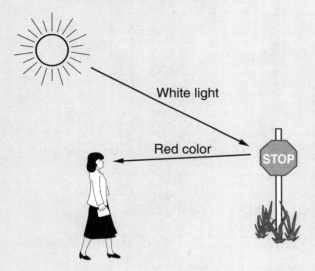

White light

Red color

STOP

(a) Describe a demonstration that shows that white light contains the visible spectrum.
(b) What happens to the orange, yellow, green, blue, and violet wavelengths that strike the stop sign?
(c) What happens to the red wavelength that strikes the stop sign?
5. Base your answers to the questions on the reading passage below and on your knowledge of science.

   Objects in space, such as the moon and Earth, receive sunlight on one side and cast a shadow toward space on the opposite side away from the sun. These objects cast shadows because they are opaque. When light strikes an opaque object, a region of darkness is produced in the opposite direction away from the light source. The shadow occupies all the space behind the object. Only when the shadow falls on another object can the shadow be seen.

(a) Explain how moonlight is produced.
(b) Several times each year, Earth's shadow is cast on the moon. Describe what happens to the moonlight when this occurs.
(c) Identify and describe the two parts of a shadow.
6. Base your answers to the questions on the diagram below and on your knowledge of science. The diagram shows four sound waves with different properties.

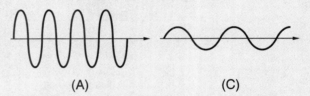

(A)   (C)

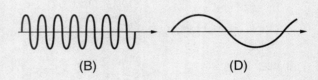

(B)   (D)

(a) Which wave has the lowest frequency and which wave has the highest frequency?
(b) Which wave has the highest pitch?
(c) Which wave produces the loudest sound?

Chapter Puzzle (*Hint:* The words in this puzzle are terms used in the chapter.)

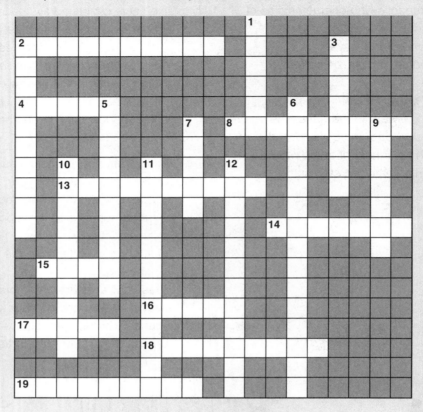

Across

2 bending of light rays caused by a change in the medium

4 visible form of radiant energy

8 height of a wave

13 light rays change into heat energy

14 effect in pitch caused by a change in distance

15 piece of transparent glass or plastic

16 energy produced by a vibrating object

17 property of sound determined by frequency

18 loudness of sound

19 number of waves that pass a fixed point in a given amount of time

Down

1 substance sound waves travel through

2 bouncing of light rays off a surface

3 visible _____: colors in the rainbow

5 type of wave that vibrates at right angles to the direction it is moving

6 type of energy that moves in waves and can travel through space

7 unit used to measure wave frequency

9 unit used to measure sound intensity

10 distance from wave crest to wave crest

11 light waves passing through an object

12 type of wave that vibrates back and forth in the direction it is moving

Chapter 7

Physical and Chemical Changes

Contents

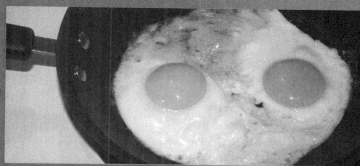

Cooking involves both chemical and physical changes.

What Is This Chapter About?

Chemistry is the study of matter and its changes. In Chapter 5, you learned about the different types of matter. In Chapter 7, we focus on how matter changes.

In this chapter you will learn:

1. Changes in matter can be physical changes or chemical changes.

2. All changes in phase are physical changes.

3. Making a mixture or separating a mixture is a physical change.

4. Solutions and suspensions are types of mixtures.

5. Stirring, heating, and crushing a solid can make it dissolve faster.

6. The amount of solute that can dissolve in a solvent depends on the temperature of the mixture.

Science in Everyday Life: Cooking

Do you have a laboratory in your home? You could say that your kitchen is a laboratory. Every day you do some of the same things in your kitchen that a chemist does in a lab. Cooking and preparing food involve many of the changes in matter discussed in this chapter. For example, preparing a pot of coffee involves making a solution and filtering it. Frying an egg or toasting a piece of bread causes chemical changes in the food. When butter melts on your toast, or water boils in your teakettle, a phase change takes place. The science of chemistry is closely related to the art of cooking.

Internet Sites:

You will find interesting information and activities on chemistry at the following Web sites:

http://www.chem4kids.com An introduction to the topics discussed this year and many more.

http://pbskids.org/zoom/activities/sci/ Chemistry activities that you can do at home and share your results with other students.

7.1

What Is the Difference Between Physical and Chemical Change?

Objectives

Identify phase changes.

Define and give examples of physical changes.

Define and give examples of chemical changes.

Terms

vapor: the gas phase of a substance that is not normally a gas at room temperature

composition: the material that something is made of

physical change: when a substance changes but does not form a new substance

mixture: the result you get when you put two or more different substances together and they do not form a new substance

chemical change: when matter changes to produce a new substance

Phase Changes

What happens to a puddle of water after the sun comes out? You have probably noticed that the water slowly disappears. What is happening to it? The liquid water is changing into water vapor. A **vapor** is the gas formed when a substance, that is not normally a gas at room temperature, changes into a gas. *Stop and Think:* What might happen to the puddle in the winter if the temperature became very cold? *Answer:* The puddle might freeze.

What is the chemical formula for the water in the puddle? You probably recall that the chemical formula for water is H_2O.

Stop and Think: What is the formula for water vapor? What is the formula for ice? *Answer:* Both water vapor and ice have the formula H_2O. Water, ice, and water vapor all have the same formula because all are the same substance. When water freezes or evaporates, it still has the formula H_2O!

Freezing and evaporation are changes in phase. Remember you learned last year that there are three phases of matter: solid, liquid, and gas. Changing any substance from one phase to another is called a *phase change.* Table 7.1-1 lists the names of the different types of phase changes and examples of each.

Table 7.1-1. Examples of Phase Changes

Phase Change	Name	Example
Liquid to gas	Boiling or evaporating	Water boils in a teapot.
Solid to liquid	Melting	Ice melts in a glass.
Gas to liquid	Condensing	Water vapor in your breath condenses on a cold day.
Liquid to solid	Freezing	Water freezes to make ice.
Solid to gas	Subliming	Dry ice turns into carbon dioxide.
Gas to solid	Depositing	Water vapor turns into ice in a freezer.

Physical Change

During a phase change, there is no change in the **composition** of the substance. Composition describes what a substance is made of. Water vapor and ice are made of the same atoms, in the same proportions, and arranged in the same way. When ice melts, solid H_2O changes to liquid H_2O, but no new substances are formed. A change in which no new substances are formed is called a **physical change**. A physical change occurs when the arrangement of the molecules changes, but the molecules themselves do not. Figure 7.1-1 shows how the arrangement of water molecules changes during evaporation. Notice that the molecules themselves do not change. *All changes of phase are physical changes.* (See Table 7.1-2.)

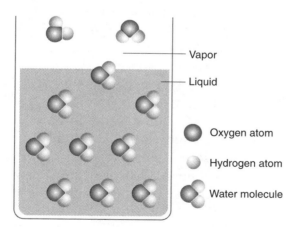

Vapor

Liquid

● Oxygen atom

○ Hydrogen atom

Water molecule

Figure 7.1-1. When water evaporates, the molecules separate, but their composition does not change. They are still H_2O.

Table 7.1-2. Examples of Physical Changes

Physical Changes
Freezing
Melting
Condensation
Boiling
Evaporation
Tearing
Crushing
Dissolving

Other Types of Physical Changes

A phase change is not the only type of physical change. When you grind or crush a solid into a powder or tear a piece of paper, you are not changing the atoms within it. The material is still composed of the same atoms, arranged in the same way. When crushing or tearing, you are breaking an object into smaller pieces, but you are not changing its composition. Crushing and tearing are also physical changes. In a chemistry laboratory, a mortar and pestle are used to crush solids into powder. (See Figure 7.1-2.) Using a mortar and pestle causes a physical change to occur.

Mixtures

What happens if we mix a solid and a liquid together? Suppose you place a teaspoonful of sand into a glass of water and stir it. What do you observe? The sand and the water do not change. They are simply mixed together in the same container. We call the resulting combination a mixture of sand and water. When two or more different substances are put together without forming new substances, they produce a **mixture**. Making a mixture is a physical change.

Suppose you place a teaspoonful of sugar into a glass of water and stir it. *Stop and Think:* How is the mixture that results different from the mixture of sand and water? *Answer:* In the mixture of sand and water, you can see both the sand and the water. In the mixture of sugar and water, the sugar is no longer visible. Is the sugar still there? You probably know that the mixture of sugar and water still tastes sweet. The sugar is still there, but we cannot see it. This special kind of mixture is called a *solution*. Solutions are discussed in Lesson 7.3. Since no new substance is formed, making a solution is still a physical change.

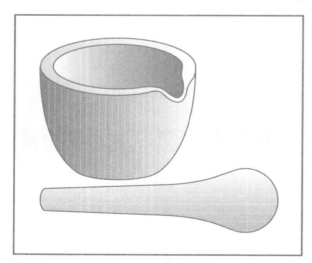

Figure 7.1-2. A pestle is used to crush solids into powder in the mortar.

Chemical Changes

Suppose you place a teaspoonful of baking soda into a glass of vinegar. The result is very different from what happened when you mixed sand or sugar with water. The baking soda eventually disappears, but before it does, you see a large amount of bubbling. What are those bubbles? The bubbles contain carbon dioxide. Carbon dioxide is a new substance that is formed when baking soda is added to vinegar. This change is not a physical change, since a new substance is formed. A change that results in the formation of a new substance is called a **chemical change** or chemical reaction.

Table 7.1-3 lists some familiar chemical changes. All changes in matter are *either* physical changes or chemical changes.

Remember that each substance is *identified* by its *properties*. When a new substance is formed, a new set of properties is observed. Carbon dioxide gas has completely different properties from those of liquid vinegar and solid baking soda. Another example of a chemical change is the reaction between sodium metal and chlorine gas. Sodium is a metal that causes an explosion when it is placed in water. Chlorine is a poisonous green gas. When sodium reacts with chlorine, the resulting substance is a white solid that you sprinkle on your french fries. The chemical change has produced a new substance, sodium chloride, also known as table salt.

Table 7.1-3. Examples of Chemical Changes

Chemical Change	Example
Burning	A match burns.
Rusting	An iron nail rusts.
Baking	A batter changes into a cake.
Spoiling of food	Milk turns sour.

Activity

You need baking soda and vinegar to do this activity. Place a small amount of baking soda in a tall glass and place the glass in the sink. Pour in some vinegar (about one-fourth of the glass). You should see the mixture begin to fizz. The reaction forms a new substance called carbon dioxide. Wash out the glass with water. Now add the same amount of baking soda to the glass. Fill the glass halfway with water and stir the mixture. Describe what happens to the baking soda. Is the baking soda still there? Test it by adding vinegar to the mixture. If you see bubbling, then you know that the baking soda is still there. Based on your results, answer the following questions:

1. What kind of change occurs when baking soda is added to water?

2. What kind of change occurs when baking soda is added to vinegar?

Explain your answers.

Interesting Facts About Phase Change

Do you like the smell of chocolate? How about freshly ground coffee beans? Have you ever thought about how the coffee gets into your nose so you can smell it? In order for something to have an odor, gas molecules must travel through the air and into your nose. Liquids must evaporate and solids sublime to form the gas molecules that cause the odor.

Dogs are much better at detecting odors than humans are. A dog's sense of smell is about 100 times better than ours. That means they need to inhale just a tiny number of molecules in order to recognize an odor. Because dogs can detect odors that humans cannot, they are used to "sniff" out bombs and drugs. When you see a police officer with a dog at an airport or sporting event, don't pet the dog or play with it. The dog is probably hard at work.

Questions

1. A chemist would describe freezing as
   (1) becoming very cold
   (2) becoming very warm
   (3) changing from a liquid to a solid
   (4) changing from a solid to a liquid

2. Which process is the *reverse* of evaporation?
   (1) boiling
   (2) condensation
   (3) freezing
   (4) sublimation

3. When alcohol evaporates, we call this a *physical* change because
   (1) the alcohol gets colder
   (2) a new substance is formed
   (3) the alcohol gets warmer
   (4) no new substance is formed

4. Melting can best be described as a
   (1) physical change because the composition changes
   (2) physical change because the composition doesn't change
   (3) chemical change because the composition changes
   (4) chemical change because the composition doesn't change

5. Which of the following is a chemical change?
   (1) dissolving sugar in water
   (2) crushing a sugar cube
   (3) melting ice
   (4) burning wood

Thinking and Analyzing

1. Mixing batter to make a cake involves only physical changes. Baking a cake, however, involves chemical change. Explain.

2. Explain why freezing water to form ice is a physical change.

7.2 How Can We Separate the Parts of a Mixture?

Objectives

Identify filtration as a method of separating the parts of a suspension.

Identify distillation as a method of separating the parts of a solution.

Terms

filtration (fihl-TRAY-shun): a method that separates the parts of a mixture based on particle size

distillation (diss-tehl-LAY-shun): a method that separates the parts of a mixture by evaporation and condensation

centrifuge (SEN-trih-fyooj): a machine that separates the parts of a mixture by spinning it

Recall that making a mixture involves only physical changes. Separating the parts of a mixture also involves only physical changes. The method you use to separate a mixture depends on the type of mixture it is. Separating a mixture of sand and water involves different steps from separating a mixture of salt and water.

Filtration

How could you separate a mixture of sand and water? A mixture of sand and water is a *suspension* in which a solid is suspended in a liquid. The pieces of sand are large enough to be visible. They are also large enough to be removed by a paper filter. (See Figure 7.2-1.) The filter in Figure 7.2-1 is folded and placed in a funnel. When the mixture is poured into the funnel, the water passes

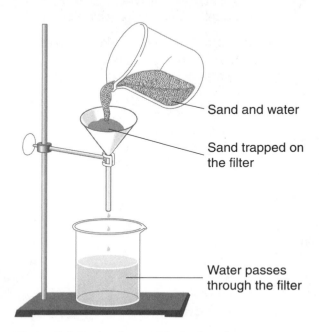

Sand and water

Sand trapped on the filter

Water passes through the filter

Figure 7.2-1. A mixture of sand and water can be separated by filtering.

through the filter but the sand does not. The tiny spaces in the filter paper are large enough to let the water molecules through, but small enough to block the larger particles of sand. This process, called **filtration**, separates the parts of a mixture based on particle size.

Evaporation

What happens to a mixture of sugar and water if it is passed through a filter? You might be surprised to find out that you cannot use a filter to separate sugar from water. Sugar water is a solution. The particles in a solution are too small to be removed by filtration. How can sugar and water be separated? When water evaporates, it leaves the sugar behind. Evaporation, however, is a very slow process. When scientists want to separate the solid from the liquid in a

solution, they often boil the mixture. Boiling speeds up the evaporation process. Boiling removes the liquid, leaving the solid behind.

Distillation

You may already know that it is not safe to drink ocean water. Ocean water is mainly a solution of salt in water. Filtering the water will not remove the salt. What might you do to make safe drinking water from ocean water? If you boil the water, you will separate the salt from the water, but will be left with just the salt. To get freshwater, we collect the water vapor produced by boiling, and cool it until it condenses back to a liquid. This two-step process of evaporation *and* condensation is called **distillation**. Figure 7.2-2 illustrates the equipment used in a laboratory distillation.

Figure 7.2-2. The equipment used in a laboratory distillation.

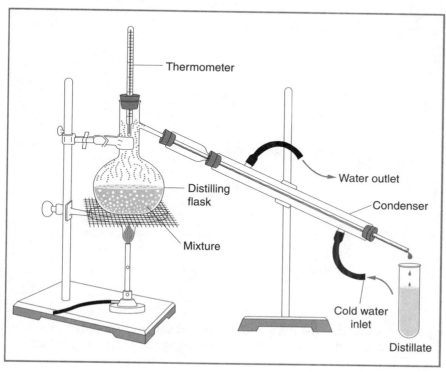

Thermometer

Distilling flask

Mixture

Water outlet

Condenser

Cold water inlet

Distillate

Other Methods of Separating a Mixture

Some methods of separation take advantage of the special properties of the substances in the mixture. *Stop and Think:* How might you separate a mixture of sand and iron powder? *Answer:* Remember that iron is attracted to a magnet. Sand is not. Iron can be separated from most other substances with a magnet.

Do you remember how to remove suspended particles from a liquid? Filtration is one method. Another method is called *centrifuging.* A **centrifuge** is a machine that spins a test tube very, very fast. When a suspension is *centrifuged*, the heavier particles move to the bottom of the test tube. A centrifuge is often used to separate the liquid part of the blood from the blood cells.

Putting It All Together

Stop and Think: How could you separate a mixture of salt and sand? Neither is attracted to a magnet. *Answer:* Take advantage of another property of salt and sand. Salt dissolves in water while sand does not. Figure 7.2-3 illustrates how you can separate a mixture of salt and sand.

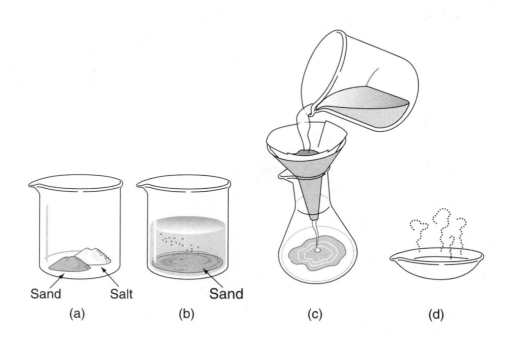

Sand   Salt   Sand

(a)   (b)   (c)   (d)

Figure 7.2-3. Separating the parts of a mixture: (a) making a mixture of salt and sand; (b) dissolving the salt in water; (c) separating the sand with a filter; and (d) evaporating the water (to get back the remaining salt).

Activity

You are going to make a solution of table salt in water and then allow the water to evaporate.

The results of this activity can vary slightly depending on the ingredients in your salt. The activity works best with pure salt. If your salt has sodium silicoaluminate (SIH-lih-koh-uh-LOO-muh-nate) or any other chemical to keep it dry, the solution formed may be slightly cloudy. This will not affect the results of the experiment.

1. Record the ingredients of your salt. Make a solution of table salt in water by placing about 1 teaspoon of salt in a half cup of warm water. Stir until all the salt dissolves. Is your solution clear or cloudy? Where did the salt go?

2. Pour the solution into a wide bowl or pan and leave it for two or three days to allow the water to evaporate. Record your findings. You may wish to bring the result to class and share your finding with your classmates. To clean the bowl after you are done, simply rinse with warm water. What happens to the salt?

Interesting Facts About Water Purification

New York State is very fortunate. It has enough freshwater for all of its people and industries. Other areas are not as lucky. California, for example, cannot meet the freshwater needs of the people living there. California has to make freshwater from ocean water. It may surprise you to find out that they do *not* use distillation to make their freshwater. Distillation is too expensive! It takes a lot of heat to boil the water on such a large scale. Burning the fuel to heat the water could also cause air pollution. Instead, another technique called *reverse osmosis (oz-MOH-sis)* is used. In reverse osmosis, water is passed through a special material, called a *membrane*, at very high pressure, separating it from the salt.

Questions

1. Why doesn't California distill its ocean water to purify it?

2. What risk to the environment can be caused by distilling ocean water to make freshwater?

Questions

1. Which of the following best separates a mixture of iron filings and black pepper?
   (1) magnet
   (2) filter paper
   (3) triple beam balance
   (4) voltmeter

2. The equipment shown at the right can be used to separate the parts of a mixture based on
   (1) boiling points
   (2) melting point
   (3) magnetism
   (4) particle size

3. A mixture of water, sugar, and sand is poured into a filter. What passes through the filter?
   (1) only the water
   (2) only the sand
   (3) water and sugar
   (4) water and sand

4. Separating a mixture by evaporating a liquid and then condensing it is called
   (1) distilling
   (2) centrifuging
   (3) filtering
   (4) dissolving

Thinking and Analyzing

1. A student is given a mixture of salt, sand, and iron filings. Describe *two* laboratory methods that the student could use to physically separate these substances.

2. Salt, pepper, and sugar are mixed with water and then filtered. What remains on the filter and what passes through it?

3. When copper sulfate is mixed with water, it produces a blue solution. When this mixture is filtered, the liquid that comes out of the filter is still blue. When the mixture is distilled, the liquid that condenses is colorless. Explain these observations.

7.3

What Are Solutions?

Objectives

Define solution.

Identify the solute and the solvent in a solution.

Terms

solution: a *mixture* in which the particles of one substance are evenly distributed throughout a second substance

solvent (SAHL-vent): in a solution, the substance that dissolves the *solute*; the solvent always keeps its phase

solute (SAHL-yoot): in a solution, the substance that dissolves in the *solvent*; the solute may change its phase

dissolve: to go into solution

soluble (SAHL-yuh-buhl): able to dissolve in a solvent

insoluble: *not* able to dissolve in a solvent

suspension (suh-SPEN-shun): a mixture that separates if allowed to stand over a period of time

Solutions

How is a mixture of sugar and water different from a mixture of sand and water? A mixture of sugar and water looks just like pure water. The sugar seems to disappear. Although we cannot see the sugar, if we taste the liquid, it tastes sweet. (**Remember—*never* taste a chemical!**) The sugar is still there, but it has been broken down into particles that are too small to be seen. A mixture in which one of the components seems to disappear is a **solution**. Recall that making a mixture is a physical change. Since a

solution is a type of mixture, making a solution is a physical change.

Parts of a Solution

In the two mixtures we just talked about, the sand eventually settles to the bottom of the glass. The sugar, however, remains spread out evenly through the water. If you took a sample of water from the top of the glass and a sample of water from the bottom of the glass, the sugar water would taste the same in both samples. A solution is a *mixture* in which the particles of one substance are evenly distributed throughout

Table 7.3-1. Examples of Solutions, Solutes, and Solvents

Example of Solution	Solute	Phase of Solute	Solvent	Phase of Solvent
Ocean	Salt	Solid	Water	Liquid
Air	Oxygen	Gas	Nitrogen	Gas
Rubbing alcohol	Water	Liquid	Isopropyl alcohol	Liquid
Brass	Zinc	Solid	Copper	Solid
Seltzer	Carbon dioxide	Gas	Water	Liquid

a second substance. Sugar and water make a solution, while sand and water do not.

In many solutions, one substance seems to disappear into the other. The substance that seems to disappear is called the **solute**. The substance that remains visible is called the **solvent**. The solute is said to **dissolve** in the solvent. When sugar is mixed with water, the sugar is the solute and the water is the solvent. Sugar dissolves in water while sand does not. We say that sugar is **soluble** in water because it dissolves in water. Sand is **insoluble** in water because it does not dissolve in water.

Types of Solutions

Most of the time, when you hear the word "solution," you probably think of a solid dissolved in a liquid. Liquids are used as solvents in a great many solutions. Water is probably the most commonly used solvent. However, solvents can also be gases or solids. Table 7.3-1 lists some examples of solutions with different kinds of solvents.

Notice that not all of the solutes in Table 7.3-1 are solids. Like the solvent, the solute can be any phase of matter. In each case, the solution that results is the same phase as the phase of the solvent.

Choosing a Solvent

Not all solutes can dissolve in all solvents. Those solutes that dissolve in water generally do not dissolve in oil. Those solutes that are soluble in oil are generally insoluble in water. When making a solution, you must carefully select your solvent to match your solute. Nail polish, for example, is insoluble in water, but is soluble in acetone. Acetone is used in many brands of nail polish remover.

Water is sometimes called the *universal solvent*. This is because it dissolves many solutes, not because it can dissolve *all* solutes.

Suspensions

Why do you shake salad dressing before pouring it on your salad? Many salad dressings contain oil and vinegar. If you mix oil and vinegar in a glass container, you will notice that the oil and vinegar separate to form two layers. The oil floats on top of the vinegar, as shown in Figure 7.3-1. We can

Figure 7.3-1. Salad dressing made from oil and vinegar forms two separate layers.

see that a mixture of oil and vinegar does not form a solution because the oil and vinegar are not evenly distributed

throughout the mixture. We say that oil is insoluble in vinegar. Shaking the salad dressing allows both the oil and vinegar to pour out at the same time. A mixture that separates is called a **suspension**. When a food bottle says on it "Shake before serving," it is telling you that the mixture is a suspension. Solutions are always evenly distributed, so they never need to be shaken!

Another difference between solutions and suspensions is that, in a suspension, the particles are large enough to be seen. Solutions are generally clear. You can see right through them. Suspensions are generally cloudy. When you buy a bottle of iced tea, some brands are clear while some are cloudy. The cloudy iced teas must be shaken because they are suspensions. The clear ones do not have to be shaken because they are solutions.

Activity

Test as many of the following solutes as you can to determine whether each is soluble in water. Place a small amount (less than a teaspoon) in a glass of water and stir. Wait a few minutes before you make your observations. If the solute truly dissolves, the resulting mixture should be transparent. The material is insoluble in water if anything settles or floats, or if the mixture turns cloudy. Make a table of your results in your notebook and share it with your classmates.

- Salt
- Pepper
- Baking soda
- Cooking oil
- Rubbing alcohol
- Baby powder
- Flour

Interesting Facts About Solubility

Grease and grime do not dissolve in water. It is difficult to remove stains from clothing with water alone. To clean clothing in water, we need to add a substance that dissolves in the water, and also dissolves grease. Soaps and detergents are chemicals that dissolve in water to form solutions that can remove grease from clothing. Soap has been used to clean clothing for thousands of years. Some fabrics, including wool, silk, and rayon, may be damaged by soap and water. Chemists have developed a different solvent to clean these fabrics.

Have you ever taken clothing to the dry cleaners? Why is the process called "dry" cleaning? You might be surprised to learn that dry cleaning uses a liquid. It is called "dry" because it does not use water. Dry cleaners use a liquid called *perc* that is a good solvent for grease and grime. You might have noticed that the clothes pick up a certain odor at the dry cleaners. Once you take the clothing out of the plastic bag, the perc evaporates, and the odor gradually disappears.

Questions

1. Which of the following materials is soluble in water?
   (1) oil
   (2) sugar
   (3) sand
   (4) grease

2. When salt and water are mixed, the salt is considered the
   (1) solvent
   (2) solution
   (3) solute
   (4) mixture

3. After a solid is added to a liquid, it is found that all of the substances are spread out evenly. This combination of substances is best described as
   (1) a mixture, but not a solution
   (2) a mixture and a solution
   (3) a solution, but not a mixture
   (4) neither a mixture nor a solution

4. A bottle of iced tea says shake well before using. After shaking it, you notice that the iced tea is cloudy. This iced tea can best be describe as a
   (1) solution
   (2) suspension
   (3) solvent
   (4) solute

5. When solid iodine is added to liquid alcohol, the resulting orange liquid is a solution called tincture of iodine. In this mixture, the
   (1) iodine is the solvent
   (2) alcohol is the solute
   (3) alcohol is the solvent
   (4) water is the solvent

Thinking and Analyzing

1. When you mix oil and water, two layers form. Is the resulting mixture a solution? Explain.

2. Jewelry that is 14 karat gold is generally 14/24 gold and 10/24 other metals, such as copper and silver. Explain why 14 karat gold is considered a solution.

7.4 What Factors Affect the Rate That a Solute Dissolves?

Objectives

Predict the effect of stirring, changing temperature, and particle size on the rate at which a solute dissolves.

Identify methods of increasing the rate at which a solute dissolves.

Terms

rate: the speed at which something happens

surface area: a measure of the amount of a substance that is exposed to other substances

The Dissolving Process

When you dissolve a solid in a liquid, the solute seems to disappear. Where does it go? Solvent molecules are attracted to solute molecules and pull them off the solid crystal. (See Figure 7.4-1.) The solvent molecules surround each of the solute molecules and carry them away to other locations in the liquid. The separated solute molecules are too small to be seen, so the solute seems to disappear.

Stirring

What do you do right after you add sugar to your tea? You probably stir the mixture. You have probably noticed that stirring makes the sugar dissolve faster. The speed at which the solute dissolves is called the **rate** of dissolving. Scientists use the word "rate" to describe how fast something happens. Stirring increases the rate of dissolving.

What other factors change the rate of dissolving? Read the following sections to find out.

Particle Size

Sugar is sold in several different forms. These include sugar cubes and granulated sugar. You probably use granulated sugar in your home. Figure 7.4-2 shows these two types of sugar. Which type dissolves faster in water? In order for a solid to dissolve, its molecules must come in contact with the solvent molecules. *Stop and Think:* When a large cube of sugar, like the one shown in Figure 7.4-2a, is placed in a glass of water, which sugar molecules are in contact with the water? *Answer:* Only the molecules on the outside of the cube are touching the water. When we pour water into a glass containing granulated sugar, the water comes in contact with much more of the sugar and dissolves it faster. When we

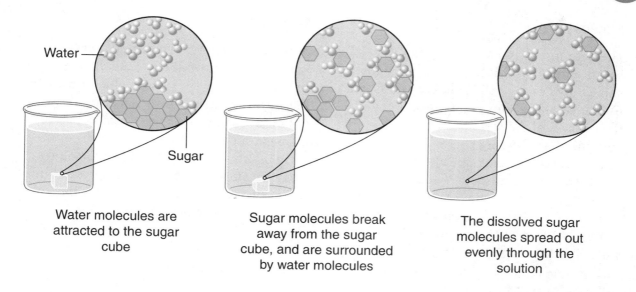

Water

Sugar

Water molecules are attracted to the sugar cube

Sugar molecules break away from the sugar cube, and are surrounded by water molecules

The dissolved sugar molecules spread out evenly through the solution

Figure 7.4-1. When a solute dissolves, its molecules are surrounded by solvent molecules and carried away.

decrease the sizes of the particles, we increase the contact between the solute and the solvent. *Decreasing* particle size *increases* the rate of dissolving.

To make solids dissolve faster, they are often crushed or ground into a powder.

Chemists use a mortar and pestle to grind solids into powder. (Look back at Figure 7.1-2 on page 208.) Powdered sugar is used in many baking recipes. It dissolves even faster than granulated sugar, because the particles are even smaller.

(a)

(b)

Figure 7.4-2. Sugar can be purchased as (a) sugar cubes and (b) granulated sugar.

Surface Area

The amount of contact between a solid solute and a liquid solvent depends on the surface area of the solid solute. **Surface area** is defined as the sum of all the areas of all the surfaces of the solid. It measures the amount of a substance that is exposed to other substances. Look at Cube A in Figure 7.4-3. The arrows point to some of the surfaces. When the cube is broken into four pieces to form Cube B, there are now more exposed surfaces. Breaking the solid into smaller and smaller pieces makes the total surface area larger and larger. *Increasing surface area increases* the rate of dissolving.

Temperature

Stop and Think: What effect do you think temperature has on the rate at which a substance dissolves? You may have observed that sugar dissolves much faster in hot water than it does in cold water. **Answer:** An *increase* in temperature of a solvent *increases* the rate at which all solid solutes dissolve. **Stop and Think:** Why do you think this is so? **Answer:** Recall that when the temperature increases, molecules move faster. The faster-moving solvent molecules hit the solid surfaces of the solute more frequently. This increase in contact between the solute and solvent molecules results in faster dissolving. In addition, the increase in temperature loosens the bonds that hold the solid together. The looser bonds allow the solid particles to break away and dissolve more quickly.

Figure 7.4-3. When particles are made smaller, there is more total surface area. When Cube A is broken into four pieces (Cube B), the original surfaces (shown with lighter arrows) remain and new surfaces (shown with darker arrows) are exposed.

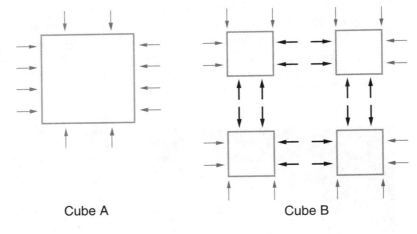

Cube A   Cube B

Interesting Facts About the Rate of Dissolving

Brewing Coffee

Many families use an electric drip coffee maker to make coffee. To make coffee, water is passed through ground coffee beans. The dissolved materials pass through a paper filter and drip into the pot below. (See figure.) Some people like their coffee strong, and some prefer it weak. By "strong" they mean that more coffee is dissolved per cup of water. What factors determine the strength of the coffee? One factor is the amount of coffee you use. Another is the amount of water. However, even using the same exact amount of coffee in the same exact amount of water will not always result in a cup of coffee of the same strength. Why?

This is a cut-away view of a drip coffee maker showing the water, filter, and coffee grinds.

Not all coffee beans are ground the same way. Coarsely ground coffee has larger particles than finely ground coffee. Which results in a stronger cup of coffee? Recall that smaller particles dissolve faster. Finely ground coffee produces a stronger cup of coffee than coarsely ground coffee if all other factors are the same.

Not all coffee makers heat the water to the same temperature. How might this affect the strength of a cup of coffee? Recall that solids dissolve faster in hotter water. A coffee maker that passes hotter water over the coffee grounds will make stronger coffee, if all other factors are the same. Even the type of filter affects the outcome. If the water passes through the filter quickly, it will spend less time in contact with the coffee, and produce a weaker cup.

When a procedure depends on so many variables, it is difficult to predict the result. Most people learn how to use their coffee maker through trial and error. Once you brew the perfect cup of coffee, remember how you did it, and keep on doing it!

Activity

Have a dissolving race! Mark a letter X on two separate sheets of paper. Pour hot water from the tap into a drinking glass until it is half full. Pour cold water from the tap into a similar glass until it is half full. Place each glass over the X. (X marks the spot!) You should be able to see the X when you look down through the water. Add one teaspoonful of sugar to each glass. Do NOT stir. Time how long it takes until one of the Xs becomes visible as you look down through the solutions. Which solution revealed the X first? Without changing the temperature, what could you do to the other solution that would reveal the X in that glass faster? Try it. Explain the results of your experiment and compare them with your classmates' results.

Questions

1. Which sample of sugar dissolves the fastest?
   (1) sugar cubes in cold tea
   (2) granulated sugar in cold tea
   (3) sugar cubes in hot tea
   (4) granulated sugar in hot tea

2. The rate of solubility of a solid in a liquid can be increased by
   (1) stirring the liquid
   (2) using larger solute particles
   (3) cooling the liquid
   (4) decreasing the surface area of the solid

3. How fast something dissolves is called the
   (1) solubility of the solute
   (2) amount of the solute
   (3) strength of the solution
   (4) rate of dissolving

4. An increase in the rate of dissolving is caused by a decrease in the
   (1) temperature
   (2) particle size
   (3) amount of stirring
   (4) surface area

Thinking and Analyzing

1. Describe three methods you can use to make salt dissolve faster in water.

2. You are making a cup of tea with a tea bag and a cup of water. Name three factors that determine the strength of the tea.

What Factors Affect the Amount of Solute That Can Dissolve in a Solvent?

Objectives

Identify the effect of changes in temperature on solubility.

Identify the effect of changes in pressure on solubility.

Interpret solubility curves.

Terms

concentrated (KAHN-sehn-tray-ted): a strong solution

concentration (KAHN-sehn-TRAY-shun): the strength of a solution

dilute (die-LOOT): a weak solution

solubility (sol-you-BILL-uh-tee): the amount of solute that can dissolve in a given solvent at a given temperature

saturated (SACH-uh-ray-ted): a solution that cannot dissolve any more solute at a given temperature

unsaturated (UHN-sach-uh-ray-ted): a solution that can dissolve more solute at a given temperature

Strong and Weak Solutions

How much sugar do you like in your tea? Do you prefer one teaspoon per cup or two? The amount of solute dissolved in a given amount of solvent is called the **concentration** of the solution. Do you like your tea strong or weak? Chemists use the terms **concentrated** and **dilute** in place of *"strong"* and *"weak"* to identify the strength of a solution. To make a solution more concentrated, we use more solute or less solvent. To make a solution more dilute, that is, to dilute a solution, we use less solute or more solvent.

What happens if you add more water to a cup of tea? The tea becomes weaker, or more dilute, because you have added more solvent. If you add more sugar to a cup of tea, the sugar becomes more concentrated, and the tea tastes sweeter. Adding more solute increases the concentration, while adding more solvent decreases the concentration. The concentration of a solution depends on the amount of solvent and the amount of solute it contains. See Figure 7.5-1, which shows how to make solutions of different concentrations.

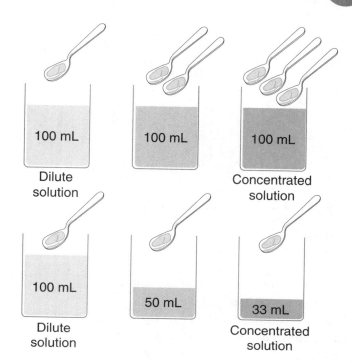

Figure 7.5-1. You can change the concentration of a solution by changing the amount of solute, or by changing the amount of solvent.

Measuring Concentration

Once you have decided how much sugar you like in your tea, how can you make sure that you always get the same result? The terms "strong" and "weak" or "concentrated" and "dilute" do not provide exact measurements. Scientists use several different units to describe exactly the concentration of a solution. One of these units is percent (%). The concentrations of many of the mixtures found in your home are expressed using percent. Although milk is not a "true" solution, percent is used to describe the concentration of fat in the milk. Milk that is 1% fat is called low-fat milk, while milk with 2% fat is often called reduced-fat milk. Percentages are used to indicate the concentrations of household solutions such as rubbing alcohol, vinegar, and hydrogen peroxide.

Solubility

What would happen if you kept adding sugar to your tea? Eventually, the added sugar would settle to the bottom and no longer dissolve. The amount of a given solid solute that can dissolve in a given solvent is limited. Once this limit is reached, any additional solute settles to the bottom and remains there. **Stop and Think:** How could you dissolve more sugar in your tea without adding more water? **Answer:** Increase the temperature of the tea! The amount of solute that can dissolve in a solvent can change if the temperature changes. Much more sugar can dissolve in hot tea than in iced tea. Most solid solutes are more soluble at higher temperatures than at lower temperatures. The maximum amount of a solute that can dissolve in a given amount of solvent is called its **solubility**. For most

Figure 7.5-2. A solution with solute on the bottom, even after stirring, is a saturated solution.

solids, as the temperature increases, the solubility increases.

A solution that contains the maximum amount of a solute it can hold at that temperature is called a **saturated** solution. A solution that contains less solute than the maximum amount is called an **unsaturated** solution. Therefore, a solution that still has undissolved solute, even after stirring, is a saturated solution. (See Figure 7.5-2.)

Solubility of Gases

The solubility of a gas in a liquid solvent also depends on the temperature of the solvent. Unlike solids, however, gases become *less* soluble when solvents are at higher temperatures, and *more* soluble when solvents are at lower temperatures. Another factor affecting the solubility of a gas is the pressure of the gas. As the pressure increases, the solubility of a gas increases. To make soda, carbon dioxide gas is dissolved in water under high pressure. This high pressure pushes more of the gas into the liquid. As long as the bottle remains sealed, the pressure remains the same. When you

look at the liquid in the bottle, you do not see any bubbles because the gas remains dissolved in the soda. When you open the bottle of soda, you decrease the pressure and can hear the gas escaping and actually see the bubbles in the soda. The bubbles that you see once the bottle has been opened are bubbles of carbon dioxide gas that are leaving the solution.

Solubility Curves

A change in temperature changes the solubility of a substance. A solubility curve is a graph that shows the solubility of a given solute in a given amount of solvent at various temperatures. Figure 7.5-3 illustrates the solubility curve for the solid potassium nitrate (KNO_3). From the graph, we can determine the maximum amount of KNO_3 that can dissolve in 100 grams of water at various temperatures. For example, at 50°C, 85 grams of KNO_3 can dissolve in 100 grams of water. What is the solubility of KNO_3 at 30°C? We see from the graph that the solubility is just less than 50 grams. We can estimate the value as 48 or 49 grams of KNO_3 per 100 grams of water.

Figure 7.5-4 shows a similar graph, which compares three different substances. The solubilities of ammonia (NH_3), potassium chlorate ($KClO_3$), and ammonium chloride (NH_4Cl) are shown. We can use the solubility curve to determine which of these solutes is most soluble at any temperature. For example, at 10°C, more grams of ammonia will dissolve in 100 grams of water than either of the other two substances. At 10°C, ammonia (NH_3) is the most soluble of these three substances and potassium chlorate ($KClO_3$) is the least

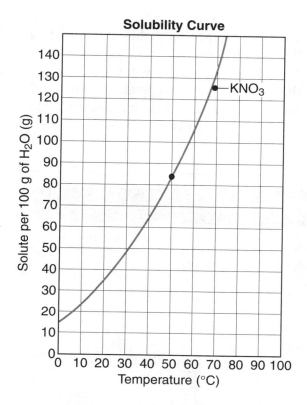

Figure 7.5-3. The solubility curve for KNO_3 shows how much solute can dissolve in 100 g of water at various temperatures.

soluble. At 90°C, ammonium chloride (NH_4Cl) is most soluble and ammonia (NH_3) is least soluble.

Comparing Solubility

We can use the solubility curves to compare the solubility of two substances at the same temperature. Compared to $KClO_3$, how many more grams of NH_4Cl will dissolve in 100 grams of water at 20°C? We see that approximately 36 grams of NH_4Cl will dissolve at that temperature, while only about 9 grams of $KClO_3$ will dissolve. To determine the *difference* in solubility, we subtract the lower solubility from the higher one: 36 grams minus 9 grams is 27 grams. Compared with $KClO_3$, 27 more grams of NH_4Cl will dissolve in 100 grams of water at 20°C.

We can also use the curves to compare the solubility of the same substance at two different temperatures. Using Figure 7.5-4, how much more soluble is NH_4Cl at 70°C than it is at 50°C? If you subtract the

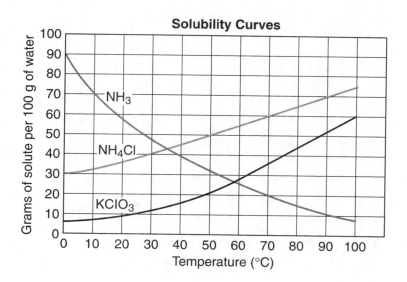

Figure 7.5-4. Solubility curves for ammonia (NH_3), ammonium chloride (NH_4Cl), and potassium chlorate ($KClO_3$).

solubility at 50°C from the solubility at 70°C, you should obtain the correct difference in solubility: 60 grams – 50 grams = 10 grams. So, how would you frame the answer to your question in words, then? "10 more grams of NH_4Cl dissolves at 70 degrees Celsius than at 50 degrees Celsius."

Notice also in Figure 7.5-4 that the NH_3 line slopes downward, while the other two lines slope upward. NH_3 is a gas, while NH_4Cl and $KClO_3$ are both solids. Recall that an increase in temperature *increases* the solubility of most solids, but *decreases* the solubility of most gases.

SKILL EXERCISE—*Interpreting a Diagram*

A series of experiments was done to determine the solubility of a solute in 100 grams of water at various temperatures. The data was recorded in the table below:

Temperature (°C)	Dissolved Solute (g)
0	74
10	78
20	84
30	90
40	98
50	106
60	116
70	128
80	141

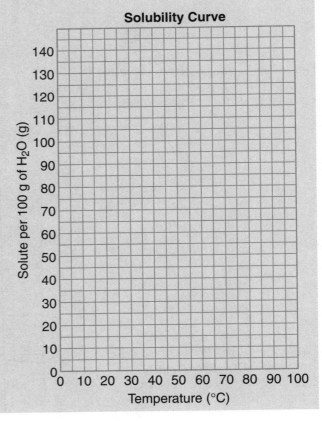

1. Copy the axes of the graph illustrated at the right onto a piece of graph paper and plot the solubility information. Place a circle around each point and connect the points with a smooth curve.

2. Predict how much solute dissolves in 100 grams of water at 55°C.

3. Is this solute a solid or a gas? Explain your answer.

Interesting Facts About Solubility

Have you ever gone scuba diving? Scuba stands for *s*elf-*c*ontained *u*nderwater *b*reathing *a*pparatus. A scuba diver is well aware of the effect of pressure on the solubility of a gas. A scuba tank contains air, which is a mixture of nitrogen and oxygen. A large amount of these gases are squeezed (compressed) into a small tank.

As a diver enters the water and swims deeper and deeper, the water pressure increases on the diver's body. These higher pressures cause more nitrogen gas to dissolve in the diver's blood. If the diver comes up too quickly, the pressure drops suddenly and nitrogen bubbles form in the diver's blood. These bubbles form in the same way that carbon dioxide bubbles form when a soda bottle is opened. In each case, a decrease in pressure causes bubbles to form. Nitrogen bubbles are very painful and even can cause death. This condition is called "*the bends.*" To prevent the bends, the diver comes up slowly, making frequent stops, called decompression stops. This allows the gas to leave the blood slowly, without forming bubbles.

Why is the condition called "the bends"? Some of the nitrogen is released into joints and muscles. When nitrogen gas enters joints, it becomes difficult and painful to move them. As a result, a person's arms and legs tend to remain in a bent position.

Questions

1. Soda manufacturers are able to dissolve large amounts of carbon dioxide gas in water by
   (1) increasing the temperature
   (2) stirring the mixture
   (3) increasing the pressure
   (4) decreasing the particle size

2. As the temperature of the mixture increases, the solubilities of most solids in liquids
   (1) increase   (3) remain the same
   (2) decrease

3. Gases are generally *most* soluble in liquids under conditions of
   (1) low temperature and low pressure
   (2) high temperature and high pressure
   (3) low temperature and high pressure
   (4) high temperature and low pressure

4. Based on Figure 7.5-4 on page 231, as the water temperature increases from 30°C to 70° C, how many more grams of $KClO_3$ will dissolve in 100 grams of water?
   (1) 11 grams   (3) 36 grams
   (2) 25 grams   (4) 47 grams

Thinking and Analyzing

1. Based on Figure 7.5-3 on page 231, describe the relationship between the temperature of water and the amount of solute that will dissolve in it.

2. When soda has lost most of its dissolved carbon dioxide, we say that it has become "flat." Why does an opened can of soda left out in a warm room become flat, while an opened can of soda in the refrigerator keeps its fizz longer?

3. A student finds two bottles labeled "Copper sulfate solution." Both solutions are blue, but one is much darker than the other. How would you describe each of these solutions?

Review Questions

Term Identification

Each question below shows two terms from Chapter 7. One of the terms is defined.
(1) Choose the term that matches the definition.
(2) Describe how the two terms are different. Following each term is the section (in parenthesis) where the description or definition of that term is found.

1. *Physical change (7.1) — Chemical change (7.1)*
   A change that produces new substances

2. *Filtration (7.2) — Distillation (7.2)*
   Separating the parts of a mixture based on particle size

3. *Solute (7.3) — Solvent (7.3)*
   The part of a solution that always keeps its phase

4. *Solution (7.3) — Suspension (7.3)*
   A mixture that separates if allowed to stand over a period of time

5. *Concentrated (7.5) — Dilute (7.5)*
   A weak solution

6. *Concentration (7.5) — Solubility (7.5)*
   The strength of a solution

Multiple Choice (Part 1)

Choose the response that best completes the sentence or answers the question.

1. Sand and iron particles that are similar in size and color are mixed together in a beaker. What method can be used to separate these particles?
   (1) Use tweezers to separate them.
   (2) Use a magnet to separate them.
   (3) Add water to the mixture.
   (4) Pour the mixture into a filter.

2. Hydrochloric acid is added to a beaker containing a piece of zinc. A change takes place, resulting in the formation of zinc chloride and hydrogen gas. This is an example of
   (1) a chemical change   (3) melting
   (2) a physical change   (4) evaporation

3. What phase change occurs when a puddle disappears over time?
   (1) melting   (3) evaporation
   (2) freezing   (4) sublimation

4. Which of the following substances is soluble in water?
   (1) oil   (3) sand
   (2) salt   (4) grease

5. Which of the following dissolves the fastest?
   (1) sugar cubes in iced tea
   (2) granulated sugar in iced tea
   (3) sugar cubes in hot tea
   (4) granulated sugar in hot tea

6. When baking soda is dissolved in water, the water is the
   (1) solute   (3) mixture
   (2) solvent   (4) solution

7. Which of the following is a physical change?
   (1) souring of milk
   (2) burning of oil
   (3) melting of ice
   (4) rusting of iron

8. A chemical change can be
   (1) mixing a sugar cube with water until it disappears.
   (2) burning a piece of charcoal until it disappears.
   (3) heating a piece of ice until it is completely changed into a liquid.
   (4) boiling a pot of water until the water completely disappears.

9. In making an omelet, which process involves a chemical change?
   (1) melting butter
   (2) chopping onions
   (3) frying eggs
   (4) adding salt

10. Which process can separate dissolved salt from water?
    (1) filtration
    (2) condensation
    (3) evaporation
    (4) centrifuging

Base your answers to questions 11 and 12 on the graph below, which shows the solubility (amount that will dissolve in 100 grams of water) of three substances at various water temperatures.

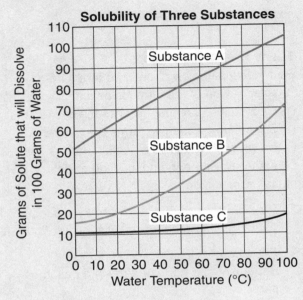

11. Which statement best describes the relationship between water temperature and solubility for the substances shown in the graph?
    (1) As water temperature increases, solubility decreases.
    (2) As water temperature increases, solubility increases.
    (3) As water temperature increases, solubility increases and then decreases.
    (4) As water temperature increases, solubility decreases and then increases.

12. As the water temperature increases from 30°C to 90°C, how many more grams of substance A will dissolve in 100 grams of water?
    (1) 30 g   (3) 50 g
    (2) 40 g   (4) 90 g

13. A mixture of salt, sand, and water is thoroughly stirred. It is then filtered through filter paper as shown in the diagram below. Which substance(s) pass through the filter?

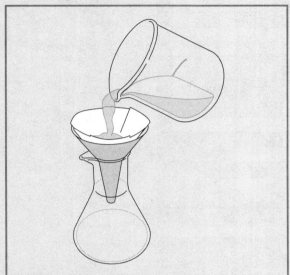

(1) only the water
(2) only the water and the sand
(3) only the water and the salt
(4) the sand, the salt, and the water

Thinking and Analyzing (Part 2)

Base your answers to questions 1 to 3 on the information below and on your knowledge of science.

A student adds a mixture of oil, sand, and salt to a beaker of water and stirs. The student stops stirring and observes that the salt is no longer visible, the oil floats to the top, and the sand sinks to the bottom of the beaker.

1. Why is the salt no longer visible after the student stops stirring?

2. Identify one way to separate the sand from the mixture in the beaker.

3. What conclusion can you come to about the solubility of oil in water?

4. Is the phase change at A in the figure below a physical change or a chemical change? Explain your answer.

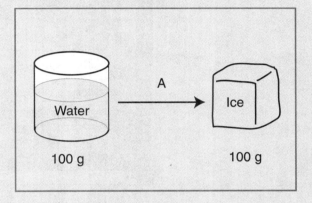

Base your answer to question 5 on the tables below, which list two elements (zinc and sulfur) and their properties. The arrows indicate that these elements may combine to form either a mixture of zinc and sulfur, or the compound zinc sulfide.

5. What evidence indicates that a chemical change took place when the zinc and sulfur combined to form zinc sulfide?

6. What three methods can you use to increase the rate at which a sugar cube dissolves in water? Explain why they work.

Element	Properties
Zinc	Good conductor of electricity, gray
Sulfur	Nonconductor of electricity, yellow

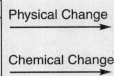

Physical Change

Chemical Change

Properties of Zinc and Sulfur Mixture
Moderate conductor of electricity, yellow and gray

Properties of Zinc Sulfide Compound
Nonconductor of electricity, white

Chapter Puzzle (*Hint:* The words in this puzzle are terms used in the chapter.)

Across

7 in a solution, the substance that dissolves in the solvent is called the _____

8 a substance that is able to dissolve in a solvent is said to be _____ in that solvent

9 a gas that is formed when a liquid evaporates

10 the amount of solute that can dissolve in a given solvent at a given temperature

11 to go into solution

12 a weak solution

15 what happens if you fight in school, or a mixture that separates if allowed to stand

16 separating the parts of a mixture through evaporation and condensation

18 the speed at which something dissolves is called the _____ of dissolving

20 a change that produces new substances is called a _____ change

21 a tool that separates the parts of a mixture by spinning the mixture very fast

22 a strong solution is also called a _____ solution

Down

1 the strength of a solution

2 separating the parts of a mixture based on particle size

3 a mixture in which the particles of one substance are evenly distributed throughout a second substance is called a _____

4 a solution that cannot dissolve any more solute at a given temperature

5 what something is made of is called its _____

6 a change that does NOT form new substances is called a _____ change

8 in a solution, the substance that dissolves the solute is called the _____

13 a substance that is NOT able to dissolve in a solvent is said to be _____ in that solvent

14 a solution that can dissolve more solute at a given temperature

17 two or more different substances put together without the formation of a new substance

19 a measure of the amount of a substance that is exposed to other substances is called its _____ area

Chapter 8

Chemistry: Understanding Chemical Reactions

Contents

Chemistry lights up the sky.

What Is This Chapter About?

In Chapter 7, you learned about physical and chemical changes in matter. Chemists use chemical equations to represent chemical changes. Chemical equations use formulas and numbers to keep track of the atoms involved in chemical changes.

In this chapter you will learn:

1. Chemical equations are used to represent chemical changes.

2. Chemical equations are balanced so that they obey the Law of Conservation of Matter.

3. Energy is involved in all chemical changes.

4. Respiration and photosynthesis illustrate how energy is involved in chemical changes.

Science in Everyday Life: Chemistry

Cars, trucks, and buses require energy to travel from place to place. These vehicles obtain energy from chemical changes that occur when gasoline or another type of fuel burns in their engines. When gasoline burns, it combines with oxygen in the air. This chemical change releases energy used to run the motor. Unfortunately, when gasoline burns it contributes to air pollution. Scientists are looking for other chemical reactions that can supply the energy we need every day without polluting the air.

Internet Site:

http://funbasedlearning.com/chemistry/ chembalancer Use this site to practice balancing equations. You need a number for every blank, so remember to fill in the number 1 when needed.

8.1

How Does a Chemical Equation Represent a Chemical Change?

Objectives

Interpret a chemical equation.

Identify the reactants and the products in a chemical equation.

Terms

word equation: a summary of a chemical reaction using words

reactants: the materials that change in a chemical reaction

products: the materials you end up with after a chemical reaction

chemical equation: a chemical reaction expressed in formulas and symbols

Law of Conservation of Matter: matter can be neither created nor destroyed during a chemical reaction

Word Equations

In Chapter 7, you reacted baking soda with vinegar. During this reaction, you observed bubbles of carbon dioxide gas. The reaction also produced water and a chemical called sodium acetate. We could summarize this chemical change using a **word equation**:

vinegar + baking soda → carbon dioxide + water + sodium acetate

We list the starting materials in a word equation to the left of the arrow. These materials are called the **reactants**. To the right of the arrow we list the new materials produced by the chemical change. The new materials are called the **products**. Everything to the left of the arrow exists before the reaction takes place. Everything to the right of the arrow exists after the reaction takes place. When reading a chemical equation, the arrow is read as "*yields*," which means "produces." A word equation shows a chemical reaction by listing the names of the reactants and products.

Chemical Equations

Recall that every substance can be represented with a chemical formula. When summarizing a chemical change, chemists often use formulas instead of names for the reactants and products. The formulas for the materials in the word equation at left are listed in Table 8.1-1. Using these formulas, the reaction between vinegar and baking soda is written as follows:

$$HC_2H_3O_2 + NaHCO_3 \rightarrow CO_2 + H_2O + NaC_2H_3O_2$$

Table 8.1-1. The Chemical Formulas for Several Substances

Name	Formula
Vinegar	$HC_2H_3O_2$
Baking soda	$NaHCO_3$
Carbon dioxide	CO_2
Water	H_2O
Sodium acetate	$NaC_2H_3O_2$
Sodium	Na
Calcium hydroxide	$Ca(OH)_2$
Chlorine	Cl_2
Sodium chloride	$NaCl$

This equation, which uses chemical formulas instead of names, is called a **chemical equation**. A chemical equation shows a chemical reaction by listing the formulas of the reactants and the formulas of the products.

Conservation of Matter

Look at the equation for the reaction between vinegar and baking soda, on page 240. Which elements appear on the reactants side? The reactant elements include hydrogen (H), oxygen (O), carbon (C), and sodium (Na). Now look at the elements on the products side. The same elements appear on the products side as on the reactants side. During a chemical reaction, atoms are rearranged, but no new atoms are formed and no atoms disappear.

How many hydrogen atoms appear on the reactants side? Remember that the number written after the symbol, called the subscript, tells us how many atoms of that element there are in the formula. A symbol that appears with no subscript represents one atom of that element. Vinegar ($HC_2H_3O_2$) contains four atoms of hydrogen. Baking soda ($NaHCO_3$) contains one atom of hydrogen. **Stop and Think:** How many atoms of hydrogen appear on the products side? **Answer:** Since there are five atoms of hydrogen on the reactants side, there must be five atoms of hydrogen on the products side. Recall that no new atoms are formed and no atoms disappear during a chemical reaction. The products are carbon dioxide (CO_2), water (H_2O), and sodium acetate ($NaC_2H_3O_2$). Notice that there are a total of five hydrogen atoms, two in H_2O and three in $NaC_2H_3O_2$.

You have just observed the **Law of Conservation of Matter**. This law states that matter can neither be created nor destroyed during a chemical reaction. We have seen that the number of hydrogen atoms does not change during our chemical reaction. Count the atoms on each side of the equation for the elements C, O, and Na. Your results should be the same as those listed in Table 8.1-2 on page 242. Table 8.1-2 shows you the symbol and number of atoms for each element on the reactants side and the products side of the equation:

$$HC_2H_3O_2 + NaHCO_3 \rightarrow CO_2 + H_2O + NaC_2H_3O_2$$

Stop and Think: Chlorine gas reacts with water according to the equation

$$Cl_2 + H_2O \rightarrow HClO + HCl$$

Count the number of atoms of each element on both sides of the equation. Does this equation obey the Law of Conservation

Table 8.1-2. Numbers of Atoms on Two Sides of a Chemical Equation

Reactants Side		Products Side	
H	5	H	5
C	3	C	3
O	5	O	5
Na	1	Na	1

of Matter? **Answer:** You found that both the reactants and the products contained 2 Cl atoms, 2 H atoms, and 1 O atom.

Parentheses in Formulas

Look at the formula for calcium hydroxide in Table 8.1-1. How many calcium, oxygen and hydrogen atoms are indicated by the formula $Ca(OH)_2$? When a formula contains a parenthesis, the subscript applies to **all** of the elements inside the parenthesis. The formula indicates that there are 1 calcium atom, 2 oxygen atoms, and 2 hydrogen atoms. A compound called aluminum sulfate has the formula $Al_2(SO_4)_3$. **Stop and Think:** How many atoms of each element are shown in this formula? **Answer:** You should find 2 aluminum atoms, 3 sulfur atoms, and 12 oxygen atoms.

SKILL EXERCISE—*Complete the Reaction*

Each of the reactions below is missing one of the reactants or products. Use the Law of Conservation of Matter to find the missing substance.

Example:

$$CaO + \underline{\quad} \rightarrow CaCO_3$$

We are missing 1 C atom and 2 O atoms on the reactants side. The missing substance is CO_2.

1. $Ca + CuCl_2 \rightarrow CaCl_2 + \underline{\quad}$
2. $PbO_2 \rightarrow \underline{\quad} + O_2$
3. $NaOH + HCl \rightarrow H_2O + \underline{\quad}$
4. $\underline{\quad} + H_2O \rightarrow NH_4OH$
5. $Zn + H_2SO_4 \rightarrow H_2 + \underline{\quad}$
6. $Ca(OH)_2 \rightarrow CaO + \underline{\quad}$

Interesting Facts About Formulas

Do you have a carbon monoxide detector in your home? Carbon monoxide is a deadly gas. It forms when fuels such as gasoline, charcoal, or cooking gas do not burn completely because of a lack of oxygen. Normally, carbon dioxide forms when fuels burn, but when the supply of oxygen is insufficient, carbon monoxide forms. The formula for carbon monoxide is CO, and the formula for carbon dioxide is CO_2. We exhale some CO_2 when we breathe, and it is found in the air around us. In small quantities, CO_2 is harmless. On the other hand, people who leave their cars running in closed garages, or try to cook indoors with charcoal may be poisoned by CO gas. What a difference an "O" makes!

Questions

1. Which statement is **true** about the reaction $CaO + CO_2 \rightarrow CaCO_3$?
(1) The only reactant is CaO.
(2) The only reactant is $CaCO_3$.
(3) Both CaO and CO_2 are reactants.
(4) CaO, CO_2, and $CaCO_3$ are all reactants.

2. The word equation for the burning of charcoal is:

   carbon + oxygen $\rightarrow$ carbon dioxide.

   The correct formula for the **product** of this reaction is
(1) C   (3) CO_2
(2) O_2 (4) H_2O

3. During a chemical reaction
(1) new atoms are created
(2) atoms are both created and destroyed
(3) atoms are destroyed
(4) atoms are neither created nor destroyed

4. Butane gas, which is used in gas lighters, has the formula C_4H_{10}. How many carbon atoms are represented by this formula?
(1) 1   (3) 14
(2) 10   (4) 4

Thinking and Analyzing

Use the following chemical equation to answer questions 1-3.

$$NaOH + HCl \rightarrow NaCl + H_2O$$

1. How many hydrogen atoms are shown on the reactants side?

2. List the products of this reaction.

3. Prove that this reaction obeys the Law of Conservation of Matter.

8.2 How Do We Balance Chemical Equations?

Objectives

Determine if a chemical equation is balanced.

Use a balanced chemical equation to illustrate the Law of the Conservation of Mass.

Use coefficients to balance chemical equations.

Terms

coefficient (koh-uh-FISH-uhnt): a number placed before a formula in a balanced chemical equation

Law of Conservation of Mass: the total mass of all the reactants is equal to the total mass of all the products

conserved (kuhn-SURVED): kept the same

Balanced Chemical Equations

One chemical change that was discussed in Lesson 7.1 was the reaction between sodium and chlorine. When sodium reacts with chlorine, they produce sodium chloride. *Stop and Think:* What is the word equation for this reaction? *Answer:* sodium + chlorine → sodium chloride.

Now, using Table 8.1-1 on page 241, we can write the chemical equation for this reaction:

$$Na + Cl_2 \rightarrow NaCl$$

We can also show this reaction by drawing circles to represent the atoms involved. (See Figure 8.2-1.)

Notice that there are 2 chlorine atoms on the reactants side (the left side) but only 1 chlorine atom on the products side (the

Key:

Cl Chlorine atom Na Sodium atom

Figure 8.2-1. A model of the chemical reaction between sodium and chlorine.

right side). What happened to the other chlorine atom? *Stop and Think:* Can matter disappear during a chemical reaction? *Answer:* Absolutely NOT! Recall that all chemical reactions must obey the Law of Conservation of Matter.

We need to include the "missing" chlorine atom in our model. The equation for the reaction tells us that the only product is NaCl. The missing chlorine atom

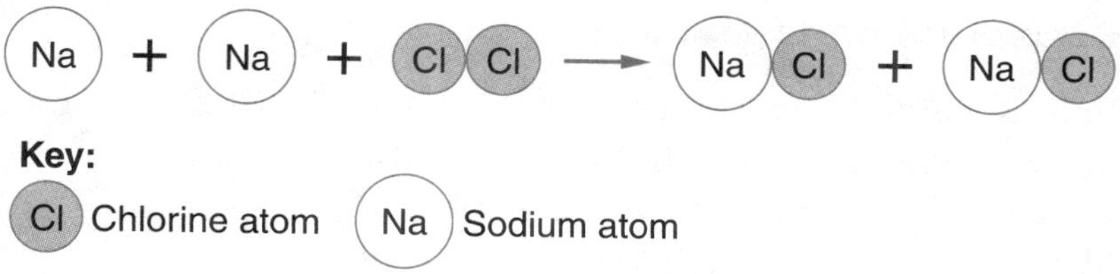

Figure 8.2-2. A model of a balanced chemical reaction between sodium and chlorine.

Na + Na + Cl Cl ⟶ Na Cl + Na Cl

Key:

Cl Chlorine atom   Na Sodium atom

must be part of a *second* NaCl molecule. Therefore, there must have been a second sodium atom on the reactants side that produced an additional NaCl on the products side. Figure 8.2-2 shows this reaction using a model that obeys the Law of Conservation of Matter.

Notice that in Figure 8.2-2 the number of sodium atoms on the left side of the arrow is the same as the number of sodium atoms on the right side of the arrow. The same is true of the chlorine atoms. How can we change our chemical equation so that it obeys the Law of Conservation of Matter? Since we are now starting with two sodium atoms, we write the number 2 in front of the symbol for sodium. In the same way, we show that we are producing 2 sodium chloride molecules:

$$2Na + Cl_2 \rightarrow 2NaCl$$

This new chemical equation is called a *balanced* chemical equation. A balanced chemical equation uses numbers called **coefficients** to indicate how many molecules and atoms of each substance are involved in the reaction. In a chemical reaction, no atoms are created, and no atoms are destroyed. The atoms are only rearranged. In a balanced equation, the number of atoms of each element in the reactants side of the equation is equal to the number of atoms of each element in the products side of the equation.

Using Coefficients

The balanced equation below shows the reaction of nitrogen (N_2) with hydrogen (H_2) to produce ammonia (NH_3):

$$3H_2 + N_2 \rightarrow 2NH_3$$

Stop and Think: How many atoms of each element appear on the reactants side of the equation?

Answer: 6 hydrogen atoms and 2 nitrogen atoms appear on the reactants side of the equation. You must multiply the coefficient by the subscripts of each atom in the formula to get the total number of atoms. If no subscript appears after an atom, its subscript number is 1. Multiplying the coefficient 3 by the subscript 2 in H_2 results in a total of 6 hydrogen atoms. You do NOT multiply the coefficient 3 by N_2 because N_2 is a separate formula. Separate formulas appear in the equation between + signs. Each formula in the equation has its own coefficient. If no number precedes a formula, the formula has a coefficient of 1. N_2 has a coefficient of 1 because no number precedes it. Look at the $2NH_3$ on the right side of the equation. Multiply the coefficient 2 by 1 N atom to get a total of 2 nitrogen atoms ($2 \times 1 = 2$). Multiply the coefficient 2 by 3 H atoms to get a total of 6 hydrogen atoms ($2 \times 3 = 6$). The number of atoms of each element is the same on both sides of the equation.

Conservation of Mass

In a chemical reaction, how does the mass of the products compare to the mass of the reactants? The Law of Conservation of Matter tells us that the number of atoms of each element cannot change. Therefore, the mass cannot change, either. We say that the mass is *conserved*, which means that it does not change. The total mass of all of the reactants is equal to the total mass of all of the products. (See Figure 8.2-3.) This statement is the **Law of Conservation of Mass**.

If we had not balanced our equation, and left it as $Na + Cl_2 \rightarrow NaCl$, mass would not have been conserved. (See Figure 8.2-4.)

Stop and Think: How is The Law of Conservation of Mass different from the Law of Conservation of Matter? *Answer:* One is used when comparing mass, and the other while comparing the number of atoms. However, if the number of atoms of each element remains the same, the *mass* must

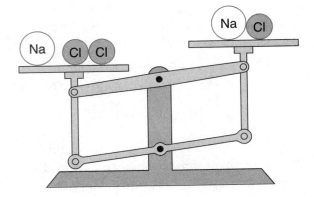

Figure 8.2-4. An unbalanced chemical equation does not obey the Law of Conservation of Mass.

also remain the same! Scientists use the two laws to express the same idea.

Finding the Coefficients

Look at the reaction between chlorine and water, shown on page 241 in Lesson 8.1.

Is this a balanced equation? You have already seen that the numbers of chlorine atoms, hydrogen atoms, and oxygen atoms are the same on both sides of the equation. This is a balanced equation. It seems that there are no coefficients in this equation. When no number appears in front of a formula, it means that the coefficient is 1.

Now, consider the reaction between sodium and chlorine,

$$Na + Cl_2 \rightarrow NaCl$$

There are more chlorine atoms on the left than on the right. The equation is not balanced. To balance it, we need to place coefficients in front of some of the formulas. How do we know what numbers to use? We look at the equation one element at a time.

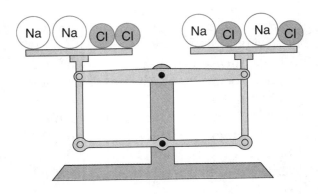

Figure 8.2-3. The Law of Conservation of Mass is illustrated by a balanced chemical equation.

There is 1 Na on the left, and 1 Na on the right. The Na atoms appear to be balanced. There are 2 Cl atoms on the left and only 1 Cl atom on the right. How can we fix this problem? We can place a coefficient of *2* in front of the NaCl. If we write "2 NaCl" it means that there are 2 Na atoms and 2 Cl atoms. The coefficient applies to the *entire* formula. Our equation now looks like this:

$$Na + Cl_2 \rightarrow 2NaCl$$

There are now 2 chlorine atoms on the left, and 2 on the right. The chlorines are balanced. The new coefficient applies to the Na of NaCl as well as to the Cl. There are now more Na atoms on the right than on the left. What to do? Simply balance the Na atoms by writing a *2* in front of the Na on the left. Our equation now looks like this:

$$2Na + Cl_2 \rightarrow 2NaCl$$

It is balanced, because there are equal numbers of sodium atoms on both sides, and equal numbers of chlorine atoms on both sides.

See if you can balance this reaction:

$$H_2 + O_2 \rightarrow H_2O$$

Which element is not balanced in the original equation? There are 2 oxygen atoms on the left, but only 1 on the right. You might be tempted just to change the subscript on the oxygen, and to write the equation: $H_2 + O_2 \rightarrow H_2O_2$. This equation is balanced, but it is a *different* reaction! The formula for water is always H_2O. Changing a subscript changes the formula for the substance. You must **NEVER** try to balance an equation by changing subscripts. How can we balance the equation correctly? We must use coefficients, not subscripts.

The first thing to do is put a 2 in front of H_2O. Now there are 4 hydrogen atoms on the right, but only 2 on the left. To make the hydrogen atoms balance, put a 2 in front of H_2. The balanced equation is:

$$2H_2 + O_2 \rightarrow 2H_2O$$

SKILL EXERCISE—*Performing an Experiment*

You can set up your own electrolysis of water. You will need a nine-volt battery, two pencils, a pencil sharpener, baking soda, water, a small glass, and a friend or relative to help out. Pencil "lead" is actually *graphite*. We use graphite because it conducts electricity.

1. Sharpen the pencils to expose the graphite at both ends. (Ask an adult to help you break off the eraser so you can sharpen that end.)

2. Place one teaspoon of baking soda in a small glass. Fill it with water and stir. (Baking soda helps the water conduct electricity; it does not take part in the reaction.)

3. Hold one pencil so that it touches the + side of the battery. Hold the other pencil (or piece of graphite) so that it touches the − side of the battery, as shown in the diagram below.

4. Place the other ends of the pencils in the water and record your observations. If no bubbles appear, make sure that the points are not broken and that each pencil makes contact with the battery.

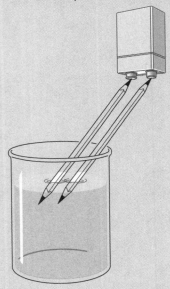

5. Compare the result for the pencil at the + side of the battery to that of the pencil at the − side of the battery.

Interesting Facts About Balanced Equations

Balancing chemical equations is not just fun, it is also important in making predictions about chemical reactions. One commonly performed chemical reaction is the electrolysis of water. This can be done using a Hoffman apparatus, shown in the figure below. A Hoffman apparatus uses electricity to split water molecules (H_2O) into hydrogen (H_2) and oxygen (O_2) molecules. Notice that the amount of gas in one tube is twice the amount in the other. How might we predict which gas is in which tube? If we balance the equation for this reaction

$$2H_2O \rightarrow 2H_2 + O_2$$

we can see that there is twice as much hydrogen produced as oxygen. A balanced equation allows us to predict the amounts of product in a chemical reaction.

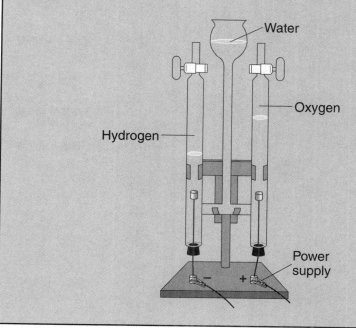

Questions

1. Which equation is correctly balanced?
   (1) $H_2O_2 \rightarrow H_2O + O_2$
   (2) $2Si + O_2 \rightarrow SiO_2$
   (3) $N_2 + 3H_2 \rightarrow 2NH_3$
   (4) $CO + O_2 \rightarrow 2CO_2$

2. What coefficient must you place in front of HCl to correctly balance the following equation?

   $$HCl + Zn \rightarrow H_2 + ZnCl_2$$

   (1) 1   (3) 3
   (2) 2   (4) 4

3. How many atoms of oxygen (O) are found on the reactants side of this balanced equation?

   $$2KClO_3 \rightarrow 2KCl + 3O_2$$

   (1) 6   (3) 3
   (2) 2   (4) 12

4. In a balanced chemical equation, the total mass of the reactants is
   (1) always greater than the total mass of the products
   (2) always equal to the total mass of the products
   (3) always less than the total mass of the products
   (4) sometimes more and sometimes less than the total mass of the products

5. The symbols below represent atoms of three different elements. Which choice shows a correctly balanced chemical equation?

Thinking and Analyzing

1. State the Law of Conservation of Mass.

2. Vinegar ($HC_2H_3O_2$) can be prepared from sodium acetate ($NaC_2H_3O_2$) using the following reaction:

   $$2NaC_2H_3O_2 + H_2SO_4 \rightarrow 2HC_2H_3O_2 + Na_2SO_4$$

   How many hydrogen atoms are represented by the term "$2NaC_2H_3O_2$"?

3. Balance the following equations by placing coefficients in the blank spaces:
   a. __NaOH + __HCl → __NaCl + __ H_2O
   b. ___Fe + ___O_2 → $2Fe_2O_3$
   c. ___N_2 + ___O_2 → ___NO_2

How Is Energy Involved in Chemical Changes?

Objectives

Identify the role of energy in photosynthesis

Identify the role of energy in respiration

State the Law of Conservation of Energy.

Describe how energy is converted from one form to another during chemical changes

Terms

photosynthesis (foe-toe-SIN-thuh-sis): the chemical process by which green plants use sunlight to convert carbon dioxide and water into glucose and oxygen

respiration (rehs-puh-RAY-shun): a chemical process in which living things use glucose and oxygen to produce carbon dioxide, water, and energy.

burning: a process in which a material reacts quickly with oxygen to produce energy

fuel (FYOOL): a substance that can be used to produce energy

Photosynthesis

You may recall that plants make their own food by a process called **photosynthesis**. During photosynthesis, plants use carbon dioxide and water to make oxygen and a sugar called glucose. The balanced chemical equation and the word equation for this process are

$$6CO_2 + 6H_2O \rightarrow C_6H_{12}O_6 + 6O_2$$

carbon dioxide + water → glucose + oxygen

Why is the reaction called photosynthesis? The prefix *photo* means "light" and the suffix *synthesis* means "putting together." During photosynthesis, green plants put together atoms from carbon dioxide and water molecules to make molecules of glucose. Light provides the energy needed to make this reaction happen. Sometimes chemists include the word "energy" in their balanced equations. Photosynthesis can be written as

$$6CO_2 + 6H_2O + energy \rightarrow C_6H_{12}O_6 + 6O_2$$

Respiration

How does your body produce the energy that keeps you alive? Your body "burns" glucose in a process called **respiration**. During respiration, glucose reacts with oxygen to produce carbon dioxide and water. The balanced chemical and word equations for this process are

$$C_6H_{12}O_6 + 6O_2 \rightarrow 6CO_2 + 6H_2O$$

glucose + oxygen → carbon dioxide + water

Where does energy appear in this reaction? Since respiration produces energy, it makes sense to list the energy along with the products. When reactions produce energy, the energy always appears on the right side of the chemical equation. Respiration can be written as

$$C_6H_{12}O_6 + 6O_2 \rightarrow 6CO_2 + 6H_2O + energy$$

You need both glucose and oxygen to supply your body with the energy that keeps you alive. You obtain glucose from the food you eat. If you do not eat, your body will "burn" stored food. Because your body can store food, you can survive for many days without eating. You obtain oxygen by breathing. If you are unable to breathe in oxygen, you will die in a matter of minutes! Your body does not store oxygen.

Burning

One way to provide the energy needed to cook a steak is to burn charcoal. The chemical equation for the **burning** of charcoal is

$$C + O_2 \rightarrow CO_2 + energy$$

A substance that can be burned to produce energy is called a **fuel**. The gas burner used in the laboratory (see Figure 8.3-1) and the gas stove in your home burn a fuel called methane. Its chemical formula is CH_4. The chemical equation for burning methane is

$$CH_4 + 2O_2 \rightarrow CO_2 + 2H_2O + energy$$

Compare the equations for burning charcoal and methane to the equation for respiration. **Stop and Think:** What do these three reactions have in common? **Answer:** In each reaction, a substance reacts with oxygen and produces energy. When a substance reacts with oxygen and produces energy, we say that the substance *burns*. We often say that we obtain our energy by burning sugar, because respiration uses oxygen to produce energy.

Energy in Chemical Reactions

Look back at the "Interesting Facts About" feature in Lesson 8.2 on page 249. In the electrolysis of water, the reaction taking place is

$$2H_2O \rightarrow 2H_2 + O_2$$

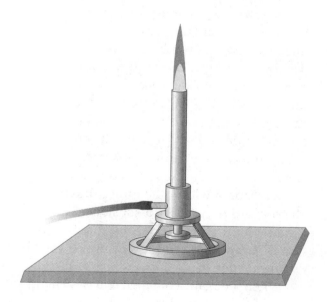

Figure 8.3-1. Burning methane in the laboratory.

How is energy involved in this reaction? The process is called *electrolysis* because it needs electrical energy to make it happen. We show the energy in this equation the same way we showed energy in the chemical equation for photosynthesis, which needed light energy to make it happen. When we include energy, the equation becomes

$$2H_2O + energy \rightarrow 2H_2 + O_2$$

All chemical reactions involve energy. Some reactions, like photosynthesis and electrolysis, absorb energy. Others, like respiration and the burning of charcoal, produce energy.

Law of Conservation of Energy

You learned that respiration produces carbon dioxide and water *from* glucose and oxygen. As predicted by the Law of Conservation of Matter, no matter is created or destroyed in this reaction. You also learned that respiration produces energy. Where did the energy come from? Like matter, energy cannot be created or destroyed during a chemical reaction. Just as there is a Law of Conservation of Matter, there is also a **Law of Conservation of Energy**. It states that energy cannot be created or destroyed. However, it can be changed from one form to another. The energy released during respiration was actually stored within the molecules of glucose and oxygen. Each substance stores a certain amount of energy. The stored energy that is involved in chemical change is called *chemical energy*.

When a chemical change releases energy, some of the chemical energy changes into heat, electricity, or some other usable form of energy. The amount of energy released is exactly equal to the amount of chemical energy used. Energy has been neither created nor destroyed. It has been changed from one form to another.

Stop and Think: Where does the chemical energy stored in food come from? **Answer:** Remember the equation for photosynthesis? During photosynthesis, energy from sunlight is stored as chemical energy in a type of sugar called glucose that is found in food.

Batteries

Where do you get the energy to run your MP3 player? You probably realize it comes from a battery. A battery is a device that converts stored chemical energy into electrical energy. The Law of Conservation of Energy also applies to batteries. There is enough stored energy in the battery to produce only a limited amount of electricity. When that limit has been reached, the battery is "dead."

Some batteries are rechargeable. When you recharge a battery, you are converting electrical energy back to chemical energy. As you use and recharge your cell phone, you are converting chemical energy to electrical, and electrical energy back to chemical. To learn more about batteries, visit the Web site: *http://www.energizer. com/learning/*

Fireworks

The energy released during chemical change is used in many applications. It runs your car, heats your home, and cooks your food. Chemical reactions also produce the gorgeous displays of heat, light, and sound we call fireworks. Fireworks involve many chemical reactions and many different substances. Each combination of substances produces a different pattern of colored light.

Red, green, and blue colors appear when the heat from chemical changes causes different elements to glow. Each element produces its own particular color.

Interesting Facts About Heat Energy

Most First Aid kits contain a device called a cold pack. A cold pack contains two chemicals that absorb heat energy when they are combined. When a cold pack is applied to an injury, it absorbs heat and keeps the injury cold. A cold pack uses a physical change, dissolving, to absorb heat. Any physical or chemical change that absorbs heat is called an *endothermic* change (*endo* means "into," and *thermic* means "heat").

Another device, often used by skiers to keep warm is called a hot pack. One type of hot pack uses oxygen from the air to cause a chemical change that releases heat. The reaction between iron in the pack and oxygen in the air is *exothermic (exo* means "out," and *thermic* means "heat"). To remember the difference between endothermic and exothermic changes, think of "endo" as "enter" and "exo" as "exit."

Questions

1. Which process absorbs energy?
   (1) photosynthesis   (3) burning wood
   (2) respiration   (4) burning food

2. The equation
   $2KClO_3 + energy \rightarrow 2KCl + 3O_2$
   shows that this reaction
   (1) releases energy
   (2) absorbs energy
   (3) produces an increase in mass
   (4) produces a decrease in mass

3. As a battery runs your MP3 player, it is converting
   (1) electrical energy into chemical energy
   (2) light energy into chemical energy
   (3) electrical energy into light energy
   (4) chemical energy into electrical energy

4. Burning can best be described as a reaction that
   (1) absorbs energy and releases oxygen
   (2) releases energy and absorbs oxygen
   (3) absorbs both energy and oxygen
   (4) releases both energy and oxygen

Thinking and Analyzing

1. A hot pack produces heat when you mix the two chemicals in it together. Describe the energy change in a hot pack.

2. Compare the energy change in photosynthesis to the energy change in respiration.

Review Questions

Term Identification

Each question below shows two terms from Chapter 8. One of the terms is defined.
(1) Choose the term that matches the definition.
(2) Describe how the two terms are different. Following each term is the section (in parenthesis) where the description or definition of that term is found

1. *Reactants (8.1) — Products (8.1)*
The materials you end up with in a chemical reaction

2. *Photosynthesis (8.3) — Respiration (8.3)*
The process in which glucose reacts with oxygen to produce carbon dioxide and water

3. *Fuel (8.3) — Burning (8.3)*
A substance that is used to produce energy

4. *Chemical equation (8.1) — Word equation (8.1)*
A chemical reaction expressed with formulas and symbols

Multiple Choice (Part 1)

Choose the response that best completes the sentence or answers the question.

1. The diagram below represents a mixture of atoms of elements A and B.

Key
● -An atom of element A
○ -An atom of element B

Which of the following diagrams below correctly shows the arrangement of these atoms after they undergo a chemical change?

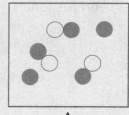

A

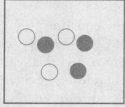

C

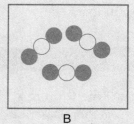

B

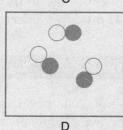

D

(1) Diagram A   (3) Diagram C
(2) Diagram B   (4) Diagram D

2. Which formula represents a reactant in the following equation?

$$2NaHCO_3 \rightarrow Na_2CO_3 + CO_2 + H_2O$$

(1) $NaHCO_3$ (3) CO_2
(2) Na_2CO_3 (4) H_2O

3. Which number is a coefficient in the following balanced chemical equation?

$$C_5H_8 + 7O_2 \rightarrow 5CO_2 + 4H_2O$$

(1) 8   (3) 6
(2) 2   (4) 7

4. In which process is oxygen used to release the energy stored in food?
(1) photosynthesis   (3) digestion
(2) respiration   (4) reproduction

5. Photosynthesis and electrolysis are alike in that they both
(1) absorb energy   (3) require sunlight
(2) release energy   (4) require electricity

6. A chemical change produces
(1) new atoms
(2) new substances
(3) an increase in mass
(4) a decrease in mass

7. How many atoms of carbon are there in the formula $HC_2H_3O_2$?
(1) 1   (3) 3
(2) 2   (4) 8

8. How many oxygen atoms appear on the reactants side in the balanced equation?

$$2Al(OH)_3 + 3H_2SO_4 \rightarrow Al_2(SO_4)_3 + 6H_2O$$

(1) 6   (3) 12
(2) 7   (4) 18

9. What is the missing formula in the following balanced equation?

$$Ca + 2HCl \rightarrow H_2 + \underline{\quad\quad}$$

(1) CaH_2 (3) $CaCl_2$
(2) $CaCl$ (4) Ca_2Cl

10. In photosynthesis, energy from sunlight is converted to a form of stored energy called
(1) chemical energy
(2) heat energy
(3) electrical energy
(4) light energy

11. Which statement about energy in chemical reactions is most accurate?
(1) All chemical reactions release energy.
(2) All chemical reactions absorb energy.
(3) Energy is never absorbed or released in a chemical reaction.
(4) Some chemical reactions absorb energy, while others release energy.

Thinking and Analyzing (Part 2)

Base your answers to questions 1–5 on the following chemical equation:

$$HC_2H_3O_2 + NaHCO_3 \rightarrow H_2O + CO_2 + NaC_2H_3O_2$$

1. How does the total mass of the products compare to the total mass of the reactants?

2. Identify the reactants in this equation.

3. How many oxygen atoms appear on the products side of this equation?

4. How many oxygen atoms appear on the reactants side of this equation?

5. Is this equation balanced? Explain how you know.

6. You need to breathe in oxygen to stay alive. What is the name of the chemical reaction that uses food and oxygen to produce energy?

7. Compare the role of energy in photosynthesis with the role of energy in respiration.

8. Balance the following equation:

$$HCl + BaBr_2 \rightarrow BaCl_2 + HBr$$

9. State the Law of Conservation of Mass.

Chapter Puzzle (*Hint:* The words in this puzzle are terms used in the chapter.)

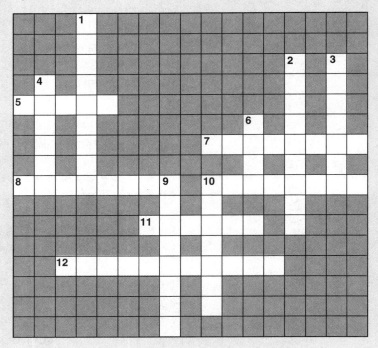

Across

5 the reactants in photosynthesis are carbon dioxide and ___

7 a summary of a chemical reaction using words is called a word ___

8 the material you end up with after a chemical reaction

10 a chemical equation that shows the law of conservation of mass must be a ___ equation

11 respiration produces water, carbon dioxide, and ___

12 the number placed before a formula in a balanced equation is called the ___

Down

1 since energy cannot be created or destroyed in a chemical reaction, we say that energy is ___

2 the materials that change in a chemical reaction

3 in respiration, ___ reacts with oxygen

4 like energy, it can neither be created nor destroyed in a chemical reaction

6 a substance that can be used to produce energy

9 the energy needed for photosynthesis comes from ___

10 reacting with oxygen to produce energy

Unit 3

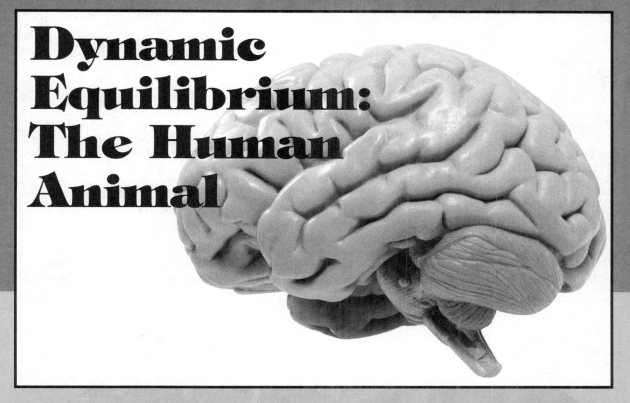

Dynamic Equilibrium: The Human Animal

Part 1—Essential Question

This unit focuses on the following essential question:

How do human body systems function to maintain homeostasis?

This unit is about you. You are an organism. Scientists call anything that is alive an organism. Earthworms, fish, grasshoppers, even tiny bacteria are organisms. All organisms are made up of building blocks called cells. You and I are made up of these cells—lots and lots of them. There are many different types of cells, all working together to keep us alive. What does it mean to be alive? A living thing must be able to obtain energy, get rid of wastes, grow, develop, respond to its surroundings, and reproduce.

The conditions that surround you are called your environment. The conditions inside an organism are called its internal environment. An organism must control its internal environment so that a proper balance of temperature, food, and wastes is maintained. If these conditions change too much, cells may not be able to do their jobs, and the organism may die. A constant internal environment within an organism is called homeostasis. All organisms must maintain homeostasis to stay alive.

The cells in large, complex organisms, such as humans, are organized into different systems. Each system has a particular job. For example, the respiratory system brings in oxygen so the body can get energy, the excretory system gets rid of wastes, and the nervous system responds to changes in the surroundings. All the body systems work together to maintain homeostasis.

Part 1—Chapter Overview

In Unit 3, you will explore how the cells of an organism are organized to make a body system. Then you will look at how several human body systems work together to maintain homeostasis.

Chapter 9 explores some of the things that all living things have in common. All living things are made up of cells and carry out certain processes called life functions. You will learn the parts of a cell, and how each of these parts keeps the cell alive. Cells are organized into tissues, organs, and organ systems. This organization allows cells to work with other cells to keep the whole organism alive.

Chapter 10 looks at many of the organs and organ systems in a very special organism—you. You will learn about the organs and organ systems in your body, and how they help your body maintain homeostasis. You will understand why you eat, why you breathe, and how the blood moves materials through your body. You will discover how the body gets rid of its wastes, moves, and responds to change. By understanding how the digestive system, respiratory system, circulatory system, excretory system, skeletal system, and nervous system function, you will gain a greater appreciation of how your body works.

Chapter 9

Levels of Organization

Contents

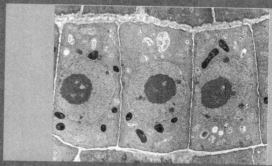

Cells are the basic units of all living things.

What Is This Chapter About?

All living things have certain common features. The most basic of these is the cell, the building block of living things. All living things carry out certain activities called life functions.

In this chapter you will learn:

1. Living things carry out *life functions* such as nutrition, respiration, and excretion.

2. The cell is the basic unit of all living things.

3. Cells carry out life functions.

4. Cells contain specialized structures to help them carry out life functions.

5. Multicellular organisms contain groups of cells working together to carry out life functions.

6. Cells are organized into tissues; tissues are organized into organs; and organs are organized into organ systems.

Science in Everyday Life: Being Alive

To stay alive, you need a constant supply of three things: food, water, and oxygen.

You need these materials to grow new cells and to provide the energy that keeps you alive. Three of your daily activities—eating, drinking, and breathing—supply your body with these materials. What you put into your body determines whether or not the cells of your body can carry out their life functions properly. Drinking plenty of water, eating healthy foods, and breathing clean air are necessary for a healthy body.

Internet Sites

You will find interesting information on being alive, the structures of cells, and the levels of organization of living things at the following Web sites:

http://www.biology4kids.com An introduction to the topics discussed this year.

http://www.cellsalive.com/cells/3dcell.htm See the many animal and plant cell organelles that we discuss in this chapter.

http://www.kidsbiology.com/biology_basics/index.php Explore three topics: What are living things?; Needs of living things; and Cells, tissues, organs, systems.

9.1 What Are Life Functions?

Objectives

Describe the cell as the basis of all living things.

Identify the processes carried out within the cell.

Identify the parts of the cell.

Relate the structures of the cell to their functions.

Terms

organism: a living thing

cell: the building block of all living things

life functions: processes that occur in all living things

homeostasis (hoe-mee-oh-STAY-sis)**:** maintaining a constant internal environment

Being Alive

You have learned that you breathe in oxygen and eat food to stay alive. You are an **organism** — a living thing. Like all living things, you require energy. This energy is supplied through the process of respiration. Where in your body does this process take place? The answer is, "Throughout your body." Your body contains trillions of living **cells**, the building blocks of all living things. Respiration takes place in each of your living cells.

Life Functions

Obtaining energy is just one of many processes that occur in all living things. These processes are called **life functions**, and are listed and described in Table 9.1-1.

Table 9.1-1. The Life Functions of Organisms

Function	Purpose
Respiration	Releases the energy stored in food
Transport	Moves materials throughout the body
Nutrition	Takes in substances and/or breaks down food into a form cells can use
Synthesis	Manufactures new molecules
Excretion	Eliminates waste materials produced by the organism
Regulation	Responds to changes in the organism's surroundings
Locomotion	Moves the organism from place to place
Reproduction	Makes more organisms of the same kind (offspring)
Growth	Increases body size, replaces damaged cells, or both
Development	Changes from a young organism into an adult

Many of these life functions help to maintain a constant set of conditions inside the body. Excretion prevents waste materials from building up in the body. Nutrition maintains the body's food supply. Transport moves materials such as food and oxygen to where they are needed. **Homeostasis** is the name given to this process of keeping the conditions inside the body constant, even though the conditions outside the body may be changing. Life functions will be discussed in greater detail throughout the rest of the book.

Cell Theory

The basic unit of all living things is called the **cell**. Cells were not discovered until the invention of the microscope in the mid-1600s. Scientists, with the development of better microscopes in the 1800s, were able to find cells in all living things. In 1838, a German scientist named Matthias Schleiden observed that all plants are made up of cells. The following year, Theodor Schwann found the same pattern in all animals. Where did these cells come from? In 1855, Rudolf Virchow stated that all cells come from previously existing cells. The work of these and other scientists led to the development of the *cell theory*.

The cell theory states that:

1. All living things are made up of cells.
2. Cells are the basic units of living things.
3. Cells come from previously existing cells.

Cells come in many sizes and shapes. Figure 9.1-1 illustrates several different types of cells that are found in the human body. These cells are specialized. They each have a particular job to do. The shape and size of each cell is related to the role of that cell.

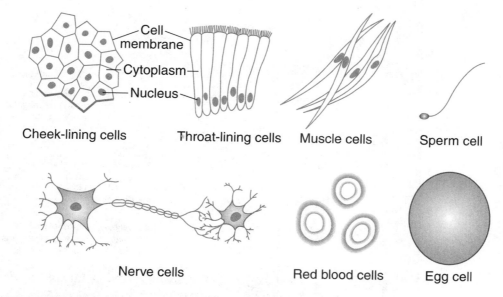

Figure 9.1-1. Several different types of human cells.

For example, nerve cells are designed to send messages through the body. These cells are long and can receive a nerve signal at one end and transmit it to another cell at the other end. Red blood cells are designed to carry oxygen. These small cells can fit through the tiniest blood vessels.

Each of the cells shown in Figure 9.1-1 contributes to a particular life function within the human body. Additionally, within each cell, life functions are taking place to keep that cell alive. The cell contains special structures that are needed to perform these life functions. Lesson 9.2 discusses these structures.

Interesting Facts About Living Things

The cell theory states that cells can come only from other living cells. Since living things are made up of cells, living things must come from other living things. You might be thinking, "Of course! Everybody knows that!" Yet, years ago, many people had a different idea.

About 300 years ago, the Italian scientist Francesco Redi wondered where maggots—small, wormlike organisms—came from. The popular belief at the time was that rotting meat turned into maggots. This idea, that living things could come from nonliving material, was called *spontaneous generation.* Redi designed an experiment to test this belief. He placed meat into two groups of jars. One group of jars was left open, and the other was sealed tightly. Diagram 1 shows what Redi observed.

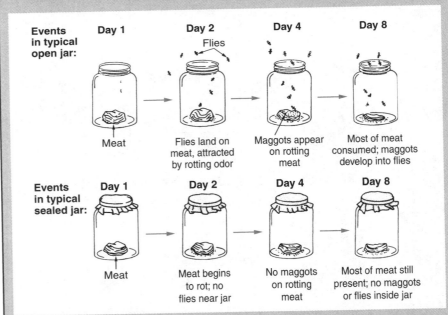

Diagram 1. Redi's first experiment: drawings show events in (top) a typical open jar; (bottom) a typical sealed jar.

Redi noted that no maggots appeared on the rotting meat in the sealed jars. However, not everyone was convinced that Redi's experiment had disproved spontaneous generation. Some people claimed that fresh air was needed for spontaneous generation to occur. Therefore, Redi performed a second experiment. He covered the jars with fine netting. The netting allowed fresh air into the jars but prevented flies from entering and landing on the meat. Diagram 2 shows what Redi observed in his second experiment.

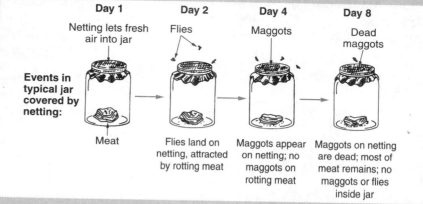

Diagram 2. Redi's second experiment.

Study both diagrams and then answer the following questions:

Questions

1. Based on Redi's experiments, where do the maggots come from?

2. What conclusion about spontaneous generation can you draw from these experiments?

Activity

Visit the following Web site:
http://www.bioscope.org/taste/history.htm
Click on the links for Hooke, Schleiden, Janssens, Schwann, and Virchow to learn more about the development of the cell theory. Then answer the following questions:

1. The first compound microscopes were made in 1590. Why weren't cells discovered until 60 years later?

2. What mistake did Schwann make in his description of cells (his *incorrect* hypothesis)?

Questions

1. The discovery of cells required the development of the
   (1) telescope   (3) microscope
   (2) balance   (4) cell theory

2. Your body produces perspiration to keep you cool in warm weather. Which life function includes this process?
   (1) nutrition   (3) reproduction
   (2) respiration   (4) regulation

3. Which life function is correctly matched with its definition?
   (1) transport—responding to changes
   (2) nutrition—increase in body size
   (3) synthesis—making new substances
   (4) regulation—supplying food for the cells

4. Which activity is most closely associated with the process of respiration?
   (1) going to the bathroom
   (2) breathing
   (3) perspiring
   (4) sleeping

Thinking and Analyzing

Read the following passage.

Your body requires additional energy when you exercise. Your cells burn more food to supply this energy. While producing this energy, your cells also produce additional carbon dioxide. Your body responds to an increase in the carbon dioxide level by breathing faster. As you breathe faster, you take in more oxygen and exhale more carbon dioxide.

1. What life process provides you with the additional energy?

2. Carbon dioxide must move from the cells to the lungs in order to leave your body. What life process moves the carbon dioxide?

3. Your body adjusts your rate of breathing in order to maintain a constant internal environment. The process that maintains constant conditions within your body is called_____.

4. All cells come from _____ _____.

9.2 How Does the Cell Carry Out Life Functions?

Objectives

Identify the three main parts of the cell.

Identify the organelles and their functions.

Distinguish between animal cells and plant cells.

Terms

nucleus (NEW-clee-uhs; plural: *nuclei*, NEW-clee-eye): the structure within the cell that controls cell activities

cell membrane: the outer covering of the cell that controls the flow of materials into and out of the cell

cytoplasm (SYE-toe-plaz-uhm): the gel-like substance that fills the cell

organelle (or-guh-NEHL): a small structure within the cytoplasm that carries out a specific life function

cell wall: the rigid outer covering of a plant cell

The Internal Structure of a Cell

The most basic cell structures are the nucleus, cell membrane, and cytoplasm. The **nucleus** is the "brain" of the cell. It controls cell activities. The nucleus contains the chemical DNA. During reproduction, DNA carries the information handed down from parent to offspring. This information tells the *ribosomes* (see Table 9.2-1) how to make proteins, important chemicals found in all cells. DNA is necessary for cell reproduction. Some cells, such as red blood cells, do not have DNA and cannot reproduce.

The **cell membrane**, the "skin" of the cell, holds the cell together. It controls the movement of materials into and out of the cell. Food, water, oxygen, and carbon dioxide are some of the materials that pass into and out of the cell through the cell membrane.

The **cytoplasm** is a gel-like substance that fills the cell. It is mainly composed of water. Within the cytoplasm are structures, called **organelles**, that perform life functions. (See Figure 9.2-1.) Each organelle is responsible for a specific life function. Table 9.2-1 lists some organelles and their functions. Notice that, except for photosynthesis, these are the *same* life functions listed in Table 9.1-1 on page 262.

Plant and Animal Cells

Look again at Figure 9.2-1. You can see that plant and animal cells share many of the same structures. However, there are a few

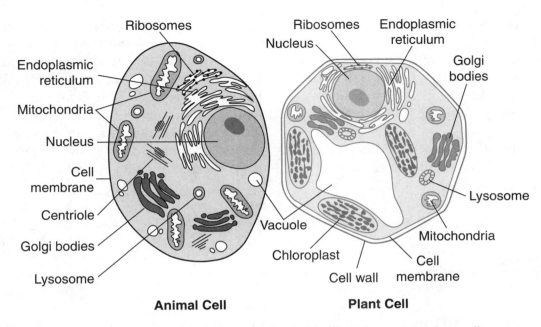

Animal Cell   **Plant Cell**

Figure 9.2-1. The internal structures of a typical animal cell and a typical plant cell.

Table 9.2-1. Some Organelles and Their Functions

Structure	Function
Mitochondria (my-toh-KAHN-dree-uh)	Respiration—food is "burned" when it combines with oxygen to produce energy. The "powerhouse" of the cell.
Ribosomes (RYE-buh-sohmz)	Synthesis—proteins are made. Ribosomes are found in the cytoplasm and on the endoplasmic reticulum (see below).
Endoplasmic reticulum (ER) (EHN-doh-PLAZ-mihk rhe-TIK-you-luhm)	Transport—move materials within the cell. The "cellular highway."
Nucleus (NEW-klee-us)	Reproduction—genetic material (DNA) is stored. The "brain" of the cell.
Vacuole (VAK-you-ohl)	Digestion and excretion—digestion occurs or excess fluid is stored. Vacuoles are usually larger in plants.
Chloroplasts (KLOR-oh-plasts)	Photosynthesis—glucose (sugar) is produced in green plants. Chloroplasts are present in plant cells but not in animals cells.
Golgi (GOHL-jee) bodies	Organelles that collect and distribute the substances produced in the cell. The cellular "post office."
Lysosomes (LYE-soh-sohms)	Contain chemicals that break down and recycle parts of the cell. Lysosomes are the "sanitation department" of the cell.

differences that make it easy to tell these cells apart. Plant cells contain an outer covering called a **cell wall**. It gives the plant cell a more rigid shape. The cell wall gives celery, carrots, and peppers their distinctive "crunchiness." Animal cells do not have cell walls.

Only plant cells contain chloroplasts. The chloroplasts contain a green chemical called *chlorophyll* (KLOR-uh-fill) that is needed for photosynthesis. Photosynthesis takes place in plants but not in animals.

Animal cells contain structures called *centrioles* that are important in duplicating DNA during reproduction. Plant cells reproduce without centrioles. Another difference you may notice is in the size of the vacuoles. Plant cells usually have much larger vacuoles than do animal cells.

Activities

1. Visit the Web site *http://www.usoe.k12.ut.us/CURR/ SCIENCE/sciber00/7th/cells/lesson/index.htm* and answer the ten questions in the quiz that follow the presentation.
2. Build your own cell! Build a model of a cell that shows at least five cell structures. Use ordinary household materials. Be creative!

Interesting Facts About Cells

Not all cells have a nucleus and organelles. The earliest forms of life were simple one-celled organisms (unicellular organisms) called prokaryotes (pro-KAH-ree-ahts). Prokaryotes have no organized nucleus. Their DNA is found in the cytoplasm. As a matter of fact, the only organelles that prokaryotes have are ribosomes. Prokaryotes include bacteria and blue-green algae.

Cells that *do* contain a nucleus and organelles are called eukaryotes (yu-KAH-ree-ahts). Some one-celled organisms and all organisms that consist of more than one cell (multicellular organisms) are eukaryotes. *Stop and Think:* Do prokaryotes carry out the same life functions as eukaryotes? *Answer:* Yes. All living things must carry out life functions. In prokaryotes, however, the life functions take place in the cytoplasm rather than in organelles.

Questions

1. Which organelles are found in plant cells but not in animal cells?
   (1) mitochondria   (3) chloroplasts
   (2) nuclei   (4) lysosomes

2. Which part of the cell controls cell activities and stores genetic material?
   (1) nucleus
   (2) cell membrane
   (3) chloroplast
   (4) golgi body

3. Mitochondria are known as the "powerhouse of the cell" because they make energy available by carrying out the life function of
   (1) excretion   (3) digestion
   (2) growth   (4) respiration

4. In humans, tubes called arteries carry blood to the cells. Which structure within the cell is most similar in function to arteries?
   (1) nucleus
   (2) endoplasmic reticulum
   (3) lysosome
   (4) golgi body

5. Which best describes the endoplasmic reticulum?
   (1) the powerhouse of the cell
   (2) the cellular highway
   (3) the sanitation department of the cell
   (4) the cellular post office

Thinking and Analyzing

For questions 1–3, name the organelle that performs each life function:

1. Respiration

2. Synthesis

3. Transport

4. Plant cells have larger vacuoles than do animal cells. Give three other differences between plant cells and animal cells.

5. Describe the functions of the cell membrane.

9.3 How Do Cells Work Together?

Objective
Identify the relationships among cells, tissues, organs, and organ systems.

Terms
unicellular (YOU-nih-CELL-yuh-ler): containing only a single cell

multicellular (MUL-tea-CELL-yuh-ler): containing many cells

tissue: a group of similar cells acting together to carry out a life process

organ: a group of tissues working together

organ system: a group of organs working together to carry out a specific life function

---

Tissues

Some organisms are made up of just one cell. These are called **unicellular** organisms. Bacteria and yeasts are examples of unicellular organisms. The single cell must carry out all the life functions that keep that organism alive.

Human beings are **multicellular**. Our bodies contain trillions of cells. Unlike the cell of a unicellular organism, the cells of a multicellular organism are specialized. Each cell has a specific task, and is usually involved in a specific life function. A nerve cell is built to send messages and a red blood cell is built to carry oxygen. (See Figure 9.1-1 on page 263.) However, one nerve cell or one red blood cell cannot do the job by itself. Many nerve cells and many red blood cells work together to carry out their assigned life function.

A group of similar cells acting together to carry out a life function forms a **tissue**. Nerve tissue sends messages through the body. Blood tissue carries materials to and from the cells. (See Figure 9.3-1.)

Table 9.3-1 lists some types of human tissues and their functions.

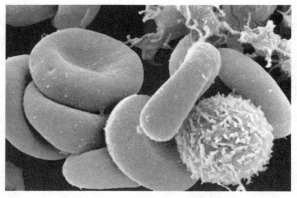

Figure 9.3-1. Blood tissue is made of red blood cells, white blood cells, and platelets.

Table 9.3-1. Types of Human Tissues and Their Functions

Tissue	Function
Blood	Transports materials throughout the body
Bone	Supports and protects the body and organs; helps the body move
Muscle	Helps the body move; aids in circulation, digestion, and respiration
Nerve	Carries messages throughout the body
Epithelial	Covers and protects the body; excretes wastes

Table 9.3-2. Important Organs and Their Functions

Organ	Function
Heart	Pumps blood throughout the body
Kidneys	Remove wastes from the blood
Lungs	Exchange gases with the environment
Stomach	Breaks down food by physical and chemical means
Brain	Controls thinking and voluntary actions
Skin	Covers and protects the body; excretes wastes

Organs

The heart pumps blood tissue through the body. The heart is composed mainly of muscle tissue, but it also contains blood tissue and nerve tissue. A group of tissues working together form an **organ**. The heart is an organ. Table 9.3-2 lists some important organs in the human body.

Like tissues, organs do not work alone. They work together with other organs to carry out a particular life function. A beating heart cannot get materials to the cells without blood and blood vessels. The heart, blood, and blood vessels together make up an organ system called the *circulatory system*. An **organ system** is a group of organs working together to carry out a specific life function.

The circulatory system carries out the function of transport, moving materials throughout the body. Table 9.3-3 lists the human organ systems and their functions.

Table 9.3-3. Human Organ Systems and Their Functions

System	Function	Example of Organs or Parts
Skeletal	Supports body; protects internal organs	Skull, ribs
Muscular	Moves organs and body parts	Arm and leg muscles
Nervous	Controls body activities; carries and interprets messages	Brain, spinal cord
Endocrine	Uses hormones to regulate body activities	Adrenal glands
Digestive	Breaks down food into a usable form	Stomach, intestines
Circulatory	Carries needed materials to body cells and waste materials away from cells	Heart, arteries, veins
Respiratory	Exchanges gases with the environment	Lungs, bronchi
Excretory	Removes wastes from the body	Kidneys, skin
Reproductive	Produces offspring	Ovaries, testes

Many of these systems will be discussed in more detail in Chapter 10.

Stop and Think: What do you call a group of organ systems working together? *Answer:* An organism!

In many ways, a large multicellular organism is like a large manufacturing company. The company has different levels of organization, from small to large. Each worker has a specific job to do within a group. Groups of workers make up a department. Different departments work together to form divisions, and divisions work together to form the company. Similarly, our bodies have levels of organization. Cells work together to form tissues, tissues work together to form organs, and organs work together to form organ systems. Organ systems work together to keep the organism alive.

Interesting Facts About Organ and Tissue Donation

What happens when an organ can no longer do its job? Fifty years ago, the failure of a vital organ, such as the heart, liver, or lung, meant certain death. Today, many lives are saved through a process called organ transplantation.

When people die, their healthy organs can be removed and donated to people who need them. This process is called organ donation. These organs save the lives of many people each year. Unfortunately, there are long waiting lists for these organs. There are over 90,000 people waiting on organ-transplant lists for kidneys, livers, hearts, lungs, and other vital organs. There are not enough organ donors available to meet the need. More than 6000 people die each year while waiting for organs that might save their lives.

Look at the figure below, which shows the back of a New York State driver's license. New York and other states are trying to make organ

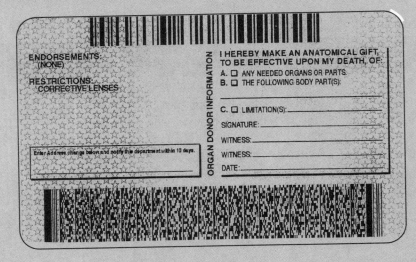

donation easier by asking people to fill out a donor card. The driver's license allows the driver to indicate that he or she is willing to be a donor.

Living people can be donors too. Red blood cells are produced in tissue called bone marrow. People who suffer from leukemia, a type of cancer of the blood or bone marrow, can sometimes be cured by a bone marrow transplant from a healthy donor. The procedure can save the life of the patient, and does not harm the donor.

Another tissue that is often donated from one living person to another is blood. When you are older, the American Red Cross might come to your school looking for blood donors. Organ and tissue donors provide the greatest gift of all, the "gift of life."

Activity

Visit the Web site
http://www.biology4kids.com/extras/video/cells_cellsystorg.html
and view the video. If you cannot view the video, you can read the transcript. Answer the following questions:
1. What is the largest organ in the body?
2. List four organs in the digestive system.
3. Which organ systems are susceptible to infection?

Questions

1. Which choice lists the levels of organization in the correct order, starting with the simplest?
   (1) cell, tissue, organ, organ system
   (2) cell, organ, organ system, tissue
   (3) tissue, cell, organ system, organ
   (4) organ, organ system, cell, tissue

2. The heart is an organ in the
   (1) respiratory system
   (2) circulatory system
   (3) excretory system
   (4) skeletal system

3. The cell illustrated to the right is used to transmit messages. The tissue that contains this type of cell is called
   (1) blood tissue   (3) muscle tissue
   (2) nerve tissue   (4) skin tissue

4. Tissue is composed of a group of
   (1) similar cells working together
   (2) organ systems working together
   (3) different organs working together
   (4) nuclei in a cell working together

5. Cells are to tissues, as tissues are to
   (1) organ systems (3) organisms
   (2) organs   (4) bacteria

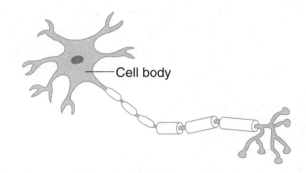

Cell body

Thinking and Analyzing

1. Which are most similar in structure: human nerve cells and human blood cells, or human nerve cells and horse nerve cells? Explain.

2. The leaf of a green plant contains several layers of different types of cells, working together to produce food through photosynthesis. Do you consider a leaf to be a tissue or an organ? Explain.

3. Describe how cells, tissues, and organs in a multicellular organism are related to one another.

4. In a multicellular organism, organ systems carry out life functions. What structures carry out life functions in a unicellular organism?

Review Questions

Term Identification

Each question below shows two terms from Chapter 9. One of the terms is defined.
(1) Choose the term that matches the definition.
(2) Describe how the two terms are different. Following each term is the section (in parenthesis) where the description or definition of that term is found.

1. *Organism (9.1) — Cell (9.1)*
   The building block of living things

2. *Nucleus (9.2) — Cell membrane (9.2)*
   The outer covering of the cell that controls the flow of materials into and out of the cell

3. *Organelle (9.2) — Cytoplasm (9.2)*
   A small structure that carries out a specific life function

4. *Tissue (9.3) — Organ (9.3)*
   A group of similar cells acting together to carry out a life process

5. *Multicellular (9.3) — Unicellular (9.3)*
   Containing only a single cell

6. *Cell wall (9.1) — Organ system (9.3)*
   The rigid outer covering of a plant cell

Multiple Choice (Part 1)

1. The diagram below shows a microscopic view of a one-celled organism. Four cell structures are labeled.

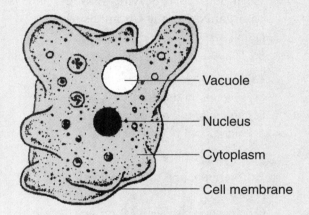

Which statement about the labeled structures is correct?

(1) They normally can be seen without magnification.
(2) They can survive outside the cell.
(3) They help carry on life activities within the cell.
(4) They cause disease within the cell.

2. Which type of tissue is most directly involved with moving materials through the body?
   (1) blood tissue   (3) muscle tissue
   (2) nerve tissue   (4) skin tissue

3. Which activity is part of the life process called nutrition?
   (1) removing wastes
   (2) eating food
   (3) breathing air
   (4) exercising

4. Cells that are designed to send and receive messages throughout the body are called
   (1) skin cells   (3) muscle cells
   (2) nerve cells   (4) blood cells

5. Complete the following analogy: Skin is to organism as membrane is to
   (1) organ   (3) cell
   (2) tissue   (4) plant

6. Cell walls are found in the cells of
   (1) animals and plants
   (2) animals, but not plants
   (3) plants, but not animals
   (4) unicellular organisms only

7. **Both** animal and plant cells contain
   (1) tissues   (3) mitochondria
   (2) chloroplasts   (4) organs

8. Which structure best completes the following analogy? Brain is to human as _____ is to cell.
   (1) membrane   (3) mitochondria
   (2) nucleus   (4) ribosome

9. Which of these parts of the body contains the greatest number of cells?
   (1) blood tissue
   (2) the heart
   (3) the arteries
   (4) the circulatory system

10. Which activity is most closely associated with the respiratory system?
    (1) breathing
    (2) moving from place to place
    (3) eating
    (4) shivering

Thinking and Analyzing (Part 2)

1. A classmate tells you that excretion is the opposite of nutrition. In what way is this statement true?

2. In many cases, organ systems found in the body have a similar function to the organelles found in cells. Copy and complete the following table by filling in the blanks. The first row has been completed for you.

Organ or Organ System	Cell Structure	Similar Function
Circulatory system	Endoplasmic reticulum	Moves materials from place to place
Respiratory system	_____	Supplies energy
Skin	Cell membrane	_____
Nervous system	_____	Controls activities

3. You accidentally touch a hot object and quickly move your hand away. Which **two** organ systems are most directly involved in your action? Explain your answer.

4. Using your knowledge of levels of organization in living things, find the missing terms for **A**, **B**, and **C** in the diagram below.

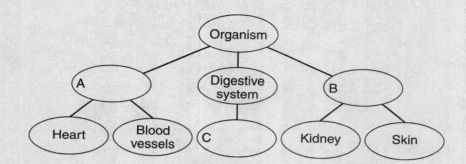

Chapter Puzzle (*Hint:* The words in this puzzle are terms used in the chapter.)

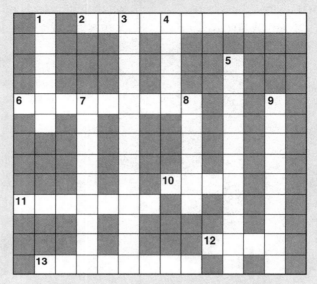

Across

2 maintaining a constant internal environment

6 processes that occur in all living things are called life ___

10 the building block of living things

11 the structure within the cell that controls cell activities

12 the rigid outer covering of a plant is called the cell ___

13 the outer covering of the cell that controls the flow of materials is called the cell ___

Down

1 a group of cells working together

3 consisting of many cells

4 a group of tissues working together

5 containing only one cell

7 the gel-like substance that fills the cell

8 a group of organs working together is called an organ ___

9 a small structure within the cytoplasm that carries out a specific life function

Chapter 10

The Human Body

Contents

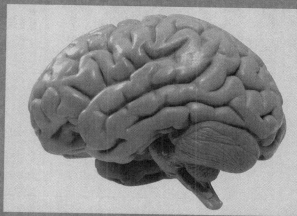

The organ in charge of homeostasis.

What Is This Chapter About?

In order to survive, all organisms must maintain homeostasis, a constant internal environment. No matter how conditions outside the organism change, conditions inside the organism must remain fairly constant.

In this chapter you will learn:

1. Many organ systems work together to maintain homeostasis.

2. The digestive system provides nutrients to the cells.

3. The respiratory system provides oxygen to the cells.

4. The circulatory system moves materials through the body.

5. The excretory system removes wastes produced by the cells.

6. The skeletal and muscular systems work together to move, protect, and support the body.

7. The nervous system and the endocrine system respond to changes.

Science in Everyday Life

There is nothing more interesting in science than learning about how your body works. This chapter explains why you eat, breathe, and go to the bathroom. Keeping your body healthy is an everyday activity.

Internet Sites:

http://www.fi.edu/biosci/index.html Take a virtual tour through the heart and the circulatory system. Click on the links in the paragraph that begins "Explore the Heart." Follow the links to other topics in this chapter.

http://health.discovery.com/tools/blausen/blausen.html Short videos show the structure and function of many different organs. They also illustrate many diseases of these organs.

10.1 How Do Humans Maintain Homeostasis?

Objectives

Explain how organ systems work together to maintain homeostasis in humans.
Describe some factors that must be kept constant through homeostasis.

Term

warm-blooded: an organism that maintains a constant body temperature

In Lesson 9.1, you learned that life functions work together to maintain homeostasis, a constant internal environment. **Stop and Think:** What do we mean by a "constant internal environment?" **Answer:** We are referring to conditions *within* our body that must be maintained. These conditions include body temperature, number of blood cells, and the amounts of glucose, oxygen, and carbon dioxide in the blood. If these conditions are not maintained, cells will not be able to carry out life functions.

Body Temperature

One factor that must remain fairly constant is your body temperature. Whatever the temperature is outside your body, your internal body temperature must always be around 37°C (98.6°F). Humans and other animals that maintain a constant body temperature are called **warm-blooded** animals. Many systems work together to maintain your body temperature.

Blood Sugar

Another factor that needs to be maintained is the amount of sugar in your blood. In a healthy person, sugar levels vary only slightly. Having too much sugar or too little sugar in the blood is dangerous. Maintaining a healthy level of sugar in your blood is the job of the endocrine system. The endocrine system produces chemicals that can increase blood-sugar levels, and chemicals that can decrease blood-sugar levels. How does sugar get into your body? You take in food, and the digestive system converts food into substances that can be carried by the blood. One of these substances is a sugar called glucose. Glucose is the fuel that keeps your body running.

Oxygen and Carbon Dioxide

In addition to glucose, blood carries many other important substances to and from cells. One of these is oxygen. Oxygen reacts with glucose to produce energy in a process called respiration. (See Chapter 9.) Cells must produce energy to stay alive, so they need a constant supply of both oxygen and glucose. When you eat, you acquire a fresh supply of glucose. Some of the glucose can be stored in our body for later use. Therefore, you do not have to eat constantly to maintain homeostasis. However, you cannot store oxygen. You need to breathe constantly to maintain the proper level of oxygen in your blood. Even a few minutes without oxygen can result in death. Obtaining oxygen is the job of the respiratory system. Delivering the oxygen to the cells is the job of the *circulatory system*. *Stop and Think:* How do your cells use carbon dioxide? *Answer:* They don't!

Carbon dioxide is a waste product of respiration. Too much carbon dioxide is dangerous. Cells transfer carbon dioxide to the blood. The blood carries carbon dioxide to the lungs, where it is exhaled. Homeostasis involves maintaining safe levels of carbon dioxide in cells and in the blood. Table 10.1-1 lists some ways that your organ systems help maintain homeostasis.

Working Together to Maintain Homeostasis

To better understand homeostasis, think about what happens when you exercise. (See Figure 10.1-1.) As you exercise, your muscles move your bones. Your hardworking muscles require more energy than usual, so your cells burn more glucose. Additional carbon dioxide enters the blood. The extra carbon dioxide is detected by your brain. Your brain sends a signal through the nervous system, causing you to breathe faster. By breathing faster, you supply your cells with more oxygen, so that they can produce more energy. In addition, the extra carbon dioxide is carried by the blood to your lungs, where it is removed and exhaled. Your heart beats faster, pushing more blood to the cells. Your hardworking body begins

Table 10.1-1. Human Systems and Homeostasis

System	Maintains homeostasis by. . .
Respiratory	Keeping oxygen and carbon dioxide at safe levels
Digestive	Providing glucose and other nutrients from food
Circulatory	Moving materials to and from cells
Nervous and endocrine	Controlling activities of the other systems
Excretory	Removing harmful materials produced by cells
Skeletal and muscular	Permitting the body to move

Figure 10.1-1. Maintaining homeostasis.

to increase in temperature. You maintain homeostasis by perspiring. When you perspire, your skin, an organ of the excretory system, produces sweat, which cools your body as it evaporates.

Just by doing a little bit of exercise, you force all of these systems, circulatory, respiratory, digestive, excretory, nervous, and endocrine to work together and maintain a constant internal environment.

Activity

You can observe how your body maintains homeostasis. While sitting in a chair, in a relaxed position, measure how many times you breathe in and out in one minute. Try to breathe naturally while making your measurements. Exercise for one minute. Some suggested exercises include: squeezing a rubber ball as fast as you can, running in place, doing push-ups, or hopping. Then sit down, and measure your breathing rate again. Did your breathing rate change? Explain your results.

Do you know how to take your pulse? If not, visit this Web site: *http://www.cimex.com/games/pulse/pulse.swf*

While seated and relaxed, measure your pulse for 30 seconds. Multiply that number by 2 to get your pulse rate in beats per minute. Exercise as before for 1 minute, sit down, and immediately measure your pulse again. What happened to your pulse rate? Explain your results.

Questions

1. Sweating maintains homeostasis by
   (1) cooling the body
   (2) removing carbon dioxide
   (3) warming the body
   (4) absorbing oxygen

2. Humans are called warm-blooded animals because
   (1) their blood temperature is always higher than room temperature
   (2) their blood temperature is at the same temperature as their surroundings
   (3) their bodies maintain a constant internal temperature
   (4) their bodies maintain a constant external temperature

3. Which organ system is correctly matched with its role in maintaining homeostasis?
   (1) muscular—removes excess carbon dioxide from the blood
   (2) respiratory—provides oxygen to the blood
   (3) excretory—takes in and breaks down food
   (4) circulatory—moves the body from place to place

4. When you exercise, your cells require more oxygen. As a result, your heart beats
   (1) faster, and you breathe slower
   (2) slower, and you breathe faster
   (3) faster, and you breathe faster
   (4) slower, and you breathe slower

5. Running is an activity that causes the cells in the muscular system to use oxygen at a faster rate. Which system responds by delivering more oxygen to these cells?
   (1) digestive   (3) circulatory
   (2) nervous   (4) excretory

Thinking and Analyzing

1. When you exercise, your cells produce more carbon dioxide. Select two organ systems and explain how they help maintain homeostasis after exercise.

10.2 Why Do We Eat?

Objectives

Identify the organs of the digestive system.

Identify different types of nutrients.

Trace the path of food through the digestive tract.

Terms

nutrient (NOO-tree-ehnt): a useful substance found in food

digestion: the process of breaking down food into a form that can be used by the cells

mechanical digestion: the physical breakdown of food into smaller pieces

chemical digestion: the process of breaking down complex substances in food into simpler substances that can be used by cells

digestive enzymes (EHN-zimes): chemicals that aid in chemical digestion

secretion: the process of producing chemicals and releasing them into the body

absorption: the movement of material into the blood

Nutrients

The mitochondria in our cells require glucose and oxygen to provide usable energy. We get oxygen by breathing. We get glucose from the food we eat. However, glucose is not the only chemical that our bodies get from food. Glucose is just one of several **nutrients**—the useful substances that are found in food. These substances include proteins, carbohydrates, fats, vitamins, and minerals. (See Table 10.2-1.) Our cells use these nutrients for energy, growth, and repair.

To learn more about vitamins and minerals, visit *http://kidshealth.org/teen /food_fitness/nutrition/vitamins_minerals.html* on the Internet.

Table 10.2-1. Some Nutrients and Their Uses

Nutrient	Use
Proteins	Supply materials for growth and repair
Carbohydrates (sugars and starches)	Provide energy
Fats and oils	Store energy
Vitamins	Prevent diseases and and play a role in life processes
Minerals	Supply materials for growth and repair and play a role in life processes

Digestion

A slice of pizza contains many of the nutrients your body needs. There are proteins and fats in the cheese, carbohydrates in the crust, and several vitamins in the tomato sauce. However, your blood is not capable of carrying a slice of pizza through your blood vessels. Foods must first be broken down by the digestive system into a form that can enter the bloodstream. **Digestion** is the process of breaking down food into a usable form.

Digestion occurs in two different ways: *mechanical digestion* and *chemical digestion.*

During **mechanical digestion**, food is physically broken down into small bits by chewing and by the action of muscles in the digestive tract.

Chemical digestion occurs when chemicals, called **digestive enzymes**, break down complex substances in the food into simpler substances. *Stop and Think:* Where do mechanical and chemical digestion begin? *Answer:* Both mechanical and chemical digestion begin in your mouth when you start chewing your food. Your teeth tear and grind the food (mechanical digestion) as enzymes in your saliva begin to break it down chemically.

The Digestive System

The digestive system, shown in Figure 10.2-1, consists of the digestive tract and accessory organs. Food travels through the body in a tube called the *digestive tract.* It begins at the mouth and continues through the *esophagus, stomach, small intestine,* and *large intestine.* The *accessory organs* are the *pancreas, gallbladder,* and *liver.* Food does

Table 10.2-2. Digestive Juices

Organ	Digestive Juice	Nutrients Acted On
Mouth	Saliva	Starches
Stomach	Gastric juice	Proteins
Small intestine	Intestinal juice	Sugars, proteins
Pancreas	Pancreatic juice	Proteins, starches, fats
Liver	Bile	Fats

Note: Bile is secreted by the liver and pancreatic juice is secreted by the pancreas into the small intestine, where digestion occurs.

not pass through the accessory organs. These organs produce digestive juices -—liquids that contain some of the digestive enzymes. These digestive juices are released into the digestive tract. The process of producing and releasing chemicals in the body is called **secretion.** (See Table 10.2-2.)

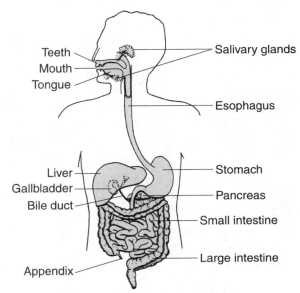

Figure 10.2-1. The human digestive system.

Table 10.2-2 lists some of the digestive juices, where they are secreted, and what nutrients they digest.

The Path of the Pizza

Let's follow the path of our slice of pizza, from the time it enters our mouth, to the time it enters our bloodstream. Mechanical digestion starts in the mouth. The teeth break the food into smaller pieces, and saliva moistens it. Chemical digestion takes place in the mouth, as enzymes in the saliva begin to break down starches.

The pizza then enters the esophagus, which connects the mouth to the stomach. Muscular walls in the esophagus push the food down to the stomach in a process called *peristalsis* (peh-ruh-STAL-sis). The stomach secretes chemicals known as gastric juices that begin the chemical digestion of proteins. The strong, muscular walls of the stomach continue the process of mechanical digestion. As these muscles contract, they churn the food, which breaks it down into still smaller pieces.

From the stomach, the food passes into the small intestine. (It is called the small intestine because it is narrow, not because it is short. It is actually about 6 meters, or 20 feet, long!) Juices from the pancreas, gallbladder, and liver are secreted into the small intestine. These juices, along with secretions from the small intestine, called intestinal juices, complete the process of chemical digestion. Once digestion has been completed, digested materials are small enough to pass through the walls of the small intestine into the bloodstream. The process by which materials pass from the small intestine to the bloodstream is called **absorption**. The nutrients that were in the slice of pizza can now be carried to the cells and used.

Some of the substances in the pizza cannot be digested. These undigested substances move into the large intestine, where water is absorbed. The remaining material then passes through the large intestine and leaves the body as solid wastes.

Activity

Water is an important nutrient. Do you get enough? Keep track of how many glasses of liquid you drink in an entire day (all of the liquids that you drink are made up mostly of water). Then use the Internet and go to this Web site:

http://kidshealth.org/kid/stay_healthy/food/water.html

1. How much water *should* you drink per day?
2. Give three reasons why water is an important part of a healthy diet.

Interesting Facts About Digestion— Heartburn

Why are we discussing heartburn in a lesson on digestion? The heart is not part of the digestive system! The heart also has nothing to do with heartburn. The discomfort that many people feel in their chest is caused by gastric juices escaping from the stomach and going into the esophagus. Gastric juices contain an enzyme called pepsin that begins the digestion of proteins. Pepsin works best in an acidic environment. Your stomach produces hydrochloric acid to create such an environment. You probably know that acids are dangerous chemicals that can damage living tissue. The walls of the stomach are coated with a thick mucus that protects its cells from the acid. As long as the acid stays in the stomach, it does not cause heartburn. Unfortunately, in some people, the gastric juices containing this acid are able to leak into the esophagus. The esophagus does not have the same protective lining that the stomach has. The burning sensation that you feel in your chest is produced when stomach acid irritates the esophagus.

What do people do to relieve occasional heartburn? There are antacids available that neutralize stomach acid, and reduce the burning sensation. However, frequent, serious heartburn can indicate a more serious condition, and should be treated by a physician.

Questions

1. Food is broken down into a usable form by the
   (1) nervous system
   (2) skeletal system
   (3) digestive system
   (4) circulatory system

2. Which group lists parts of the digestive system?
   (1) heart, lungs, pituitary gland
   (2) glucose, oxygen, carbon dioxide
   (3) skin, kidneys, lungs
   (4) stomach, intestines, pancreas

3. After it leaves the mouth, food next enters the
   (1) stomach   (3) small intestine
   (2) esophagus   (4) large intestine

4. The main function of the human digestive system is to
   (1) break down food for absorption into the blood
   (2) exchange oxygen and carbon dioxide in the lungs
   (3) release energy from sugars within the cells
   (4) carry nutrients to all parts of the body

5. Which of the following is an accessory organ of the digestive system?
   (1) mouth
   (2) pancreas
   (3) stomach
   (4) small intestine

Thinking and Analyzing

The diagram below shows several organs in the human digestive system.

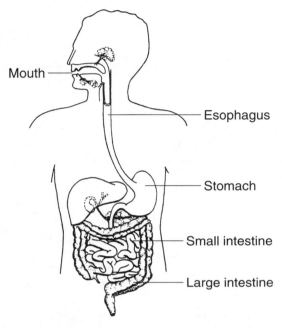

Mouth

Esophagus

Stomach

Small intestine

Large intestine

1. Explain *two* ways that food is changed as it passes through the digestive system.

2. Name two organs that participate in digestion, but are *not* labeled in the diagram.

3. What is the main function of the digestive system?

Base your answers to questions 4 through 7 on the passage below and on your knowledge of science.

We all need some fat and oil in our diet. However, not all fats and oils are alike. Fats and oils are similar in structure. They are both classified as *lipids*. Fats are solid lipids while oils are liquid lipids. There are several different types of lipids, but one group in particular, *trans fats*, has been getting a lot of attention lately. Trans fats are one of the most dangerous types of fats and have been associated with several health problems, including high cholesterol and heart disease. New York City has banned the use of trans fats in restaurants. Restaurants are now using healthier types of fats and oils for cooking foods like French fried potatoes.

4. Which statement about fats is true?
    (1) There is only one type of fat.
    (2) Nobody needs fat in his or her diet.
    (3) All oils are solid.
    (4) Trans fats are unhealthy.

5. What is the difference between fats and oils?

6. Why has New York City banned the use of trans fats?

7. What would be a good title for this reading passage?

10.3 Why Do We Breathe?

Objectives

Identify the organs of the respiratory system.

Explain the role of capillaries during the exchange of gases.

Explain how breathing occurs.

Terms

inhale: to breathe in

alveoli (al-vee-OH-lie): tiny air sacs in the lungs

breathing: moving air into and out of the lungs

diaphragm (DIE–uh-frahm): a muscle in the chest that controls breathing

exhale: to breathe out

respiration: the process of taking in oxygen, using the oxygen to provide energy, and removing carbon dioxide

cellular respiration: the chemical reaction in the cells that combines glucose and oxygen to produce energy and form carbon dioxide

---

As you know, to stay alive our cells need glucose and oxygen. The digestive system supplies the glucose. It passes from the small intestine into the bloodstream, which carries it to each cell of the body. Without oxygen, however, glucose cannot supply the energy that keeps us alive. It is the job of the respiratory system to take in oxygen and supply it to the bloodstream.

The Respiratory System

The *respiratory system,* illustrated in Figure 10.3-1, brings oxygen from the air to the blood, and returns carbon dioxide from the blood to the air. *Stop and Think:* How does oxygen get from the air into the blood? *Answer:* Through the lungs!

When you breathe in, or **inhale**, air enters your nose or mouth and passes through the *trachea,* or windpipe. The trachea branches off to each lung through tubes called *bronchi.* Smaller tubes called *bronchioles* connect the bronchi to millions of tiny *air sacs* called **alveoli**. Each air sac is surrounded by blood vessels called *capillaries.* (See Figure 10.3-2.)

Breathing

What happens to your chest as you take a deep breath? Your chest expands, increasing

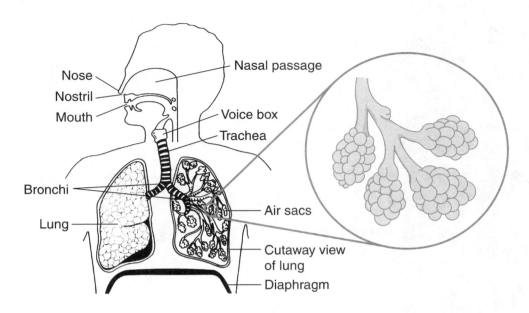

Figure 10.3-1. The human respiratory system showing an enlargement of the air sacs.

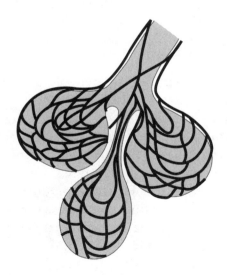

Figure 10.3-2. The air sacs in the lungs are surrounded by tiny blood vessels called capillaries.

the size of your chest cavity—the space inside your chest. While this is happening, a muscle in your lower chest cavity above your stomach, called the **diaphragm**, contracts (becomes smaller), creating additional space. (See Figure 10.3-3a on page 294.) Air rushes into your lungs to fill the space. When the diaphragm relaxes, and the chest contracts, the air is squeezed out of your lungs, causing you to breathe out. (See Figure 10.3-3b on page 294.) **Breathing** is the process of moving air into and out of the lungs.

Exchange of Gases

When you inhale, your lungs fill with fresh air that is rich in oxygen. Oxygen passes from the air sacs through the walls of the capillaries and into the blood. At the same time, carbon dioxide passes from the blood, through the capillary walls, and into the air

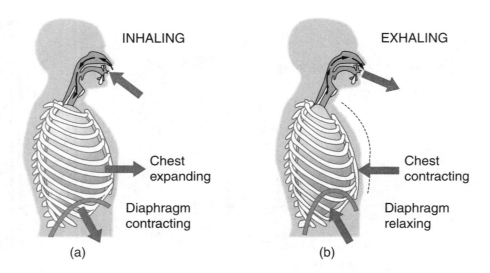

INHALING

EXHALING

Chest
expanding

Chest
contracting

Diaphragm
contracting

Diaphragm
relaxing

(a)

(b)

Figure 10.3-3. The diaphragm and chest control breathing.

sacs. The carbon dioxide is then removed from the lungs when you **exhale**, or breathe out.

Oxygen that enters the blood is carried to the cells of your body, where an exchange of gases takes place again. This time, oxygen leaves the blood and enters the cells, while carbon dioxide leaves the cells and enters the blood. The blood returns the carbon dioxide to the lungs to be exhaled. This process is repeated constantly.

Respiration—A Summary

Respiration is a three-part process. It includes:

1. The exchange of gases between the lungs and the blood

2. Using oxygen to produce energy in the cells

3. The exchange of gases between the cells and the blood

Stop and Think: Where have you seen the term "respiration" in an earlier chapter? *Answer:* In Chapters 8 and 9, you learned that respiration is a chemical reaction that takes place in living cells, and produces energy. You now know that we use the same term for the three-part process that occurs in our bodies. To avoid confusion, think of the chemical reaction within the cell that produces energy as **cellular respiration.**

Interesting Facts About Air Pressure

Most of us take breathing for granted. We take thousands of breaths each day without even thinking about it. However, for some people, breathing can become difficult and painful. They suffer from a disease called asthma (AZ-muh).

Most people who have asthma can breathe normally much of the time. However, sometimes they suffer an *asthma attack*. During an asthma attack, breathing becomes difficult, as the bronchi become swollen, limiting the amount of air that can pass through to the air sacs. Asthma attacks may occur at any time, but they are usually "triggered" by a particular condition. Some people suffer from asthma because of chemicals in the environment. Pollen from plants, tobacco smoke, and air pollution cause asthma attacks in many people. Attacks may also be triggered by cold weather, or even physical exertion.

Scientists still do not know why some people develop asthma, but they have developed several ways of treating the disease. Most asthma sufferers carry an inhaler with them. The drugs in the inhaler *relieve* the symptoms of asthma once an attack occurs. Other drugs are designed to *control* the disease by preventing the attacks from occurring. If you suffer from asthma, you may be using medications that both relieve and control the condition. Still, the best way to prevent an asthma attack may be to avoid the conditions that trigger it as much as possible.

Watch the video at *http://www.kidshealth.org/misc/movie/cc/how-asthma-affects.html* to see what happens to the bronchi and bronchioles of asthma sufferers.

Activity

How much air can you breathe out in one breath? The amount varies from person to person. The maximum amount of air you can exhale is called your *vital lung capacity*. You can measure it with a simple experiment. You will need a 1-gallon milk jug or 2-liter soda bottle and its cap, a rubber tube or a bendable straw, a large measuring cup, and a large basin.

1. Fill the bottle or jug with water, and cap it.
2. Place the basin in the sink, and fill it with water. Keep the basin in the sink during the experiment; it will overflow!
3. Invert the jug or bottle, and place it in the basin, so that the cover is below the surface of the water. (See the diagram below.)
4. Remove the cap without lifting the jug out of the water.
5. Place your rubber tube or straw so that one end is completely in the jug, as shown in the diagram. The other end must extend beyond the basin.
6. Take a deep breath, place the straw or tube in your mouth, and exhale. Keep on exhaling until you can't exhale any more, and then remove the tube from your mouth.
7. Place the cap back on the jug while the jug is still upside down in the basin.
8. Remove the jug and turn it right-side-up.
9. Uncover the jug. Use a measuring cup to measure the amount of water needed to refill the jug. This volume of water is equal to the volume of air you exhaled!

Report your lung capacity to your teacher, and compare your results with those of other students in your class. Compare the lung capacities of boys with those of girls, or of athletes with non-athletes.

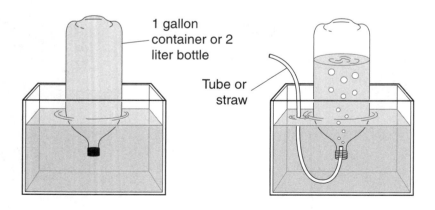

1 gallon container or 2 liter bottle

Tube or straw

Questions

1. The respiratory system includes the
   (1) heart, liver, lungs
   (2) lungs, trachea, nose
   (3) stomach, esophagus, liver
   (4) heart, arteries, veins

2. Which represents the correct order in which oxygen enters the body?
   (1) nose, trachea, bronchi, lungs
   (2) bronchi, nose, trachea, lungs
   (3) lungs, bronchi, trachea, nose
   (4) nose, bronchi, trachea, lungs

3. The exchange of gases between the air and the blood takes place in the
   (1) nose   (3) bronchi
   (2) trachea   (4) air sacs

4. When you inhale, the diaphragm
   (1) contracts, and air rushes into the lungs
   (2) contracts, and air rushes out of the lungs
   (3) relaxes, and air rushes into the lungs
   (4) relaxes, and air rushes out of the lungs

5. One purpose of respiration is to supply the cells with
   (1) carbon dioxide   (3) oxygen
   (2) nutrients   (4) water

6. At each body cell,
   (1) carbon dioxide enters the blood, and oxygen leaves the blood
   (2) both carbon dioxide and oxygen enter the blood
   (3) both carbon dioxide and oxygen leave the blood
   (4) oxygen enters the blood, and carbon dioxide leaves the blood

Thinking and Analyzing

1. Describe the exchange of gases that occurs between the air sacs and the capillaries.

2. We breathe to provide our cells with oxygen for cellular respiration. The steps involved in this process are listed below, but their order has been jumbled. Place these steps in the correct order in your notebook.
   A—Sugar is burned to make carbon dioxide.
   B—Air enters the air sacs.
   C—Oxygen moves into the body cells.
   D—The diaphragm contracts.
   E—Carbon dioxide moves to the lungs.
   F—Oxygen enters the blood.
   G—Carbon dioxide enters the blood.
   H—Carbon dioxide leaves the blood.

3. What is the difference between breathing and respiration?

10.4 How Are Materials Transported Through the Body?

Objectives

Identify the organs of the circulatory system.

Explain the differences among red blood cells, white blood cells, and platelets.

Identify the roles of arteries, veins, and capillaries.

Terms

hemoglobin (HEE-muh-gloh-bin): a protein in red blood cells that carries oxygen

infection: the illness or damage to cells that is caused by bacteria, viruses, or other microscopic organisms

platelet (PLAYT-lit): a blood cell that causes blood to clot

plasma (PLAZ-muh): the liquid part of the blood

atria (AY-tree-uh): upper chambers of the heart that receive blood

ventricles (VEHN-trih-kuhlz): lower chambers of the heart that pump the blood

arteries (AHR-tuh-rhees): blood vessels that carry blood away from the heart

veins: blood vessels that carry blood toward the heart

capillaries (CAP-uh-lehr-ees): tiny blood vessels that connect arteries to veins

lymph (LIMF): a fluid that surrounds the body cells

The Circulatory System

Nutrients provided by the digestive system and oxygen provided by the respiratory system must be transported to all body cells. The cells produce waste products that must be carried away from the cells. It is the job of the *circulatory system* to move materials through the body.

The circulatory system contains the *blood, heart, blood vessels* (*arteries, veins,* and *capillaries*), *lymph,* and *lymph vessels.*

Blood is a liquid tissue that contains red blood cells, white blood cells, and platelets.

(See Figure 9.3-1 on page 272.) Blood also carries dissolved nutrients, wastes, and hormones. (Hormones are the chemicals produced by endocrine glands. They will be discussed in more detail in Lesson 10.7)

Red Blood Cells

As you know, blood carries oxygen to the cells. Oxygen is carried by the red blood cells. Red blood cells contain **hemoglobin**, a protein that contains iron. Hemoglobin can bond with oxygen atoms and bring them to the cells. Hemoglobin is responsible for the red color of blood. When hemoglobin absorbs more oxygen, it turns a brighter red. Blood that is lacking oxygen is a duller shade of red. (Blood is never blue!)

White Blood Cells

White blood cells are your body's defense against infections. An **infection** occurs when viruses, bacteria, or other microscopic organisms enter your body and make you ill. White blood cells attack the invading organisms. When healthy people get sick with the flu, a cold, or a bacterial infection, they generally recover in a few days. White blood cells first recognize the invaders, and then attack and destroy them. An increase in the number of white blood cells in the blood may indicate that your body is fighting an infection.

Platelets

Have you ever seen your own blood? At one time or another, you have probably skinned your knee or had a bloody nose. What did you do to stop the bleeding? If it was a small wound, you probably did nothing at all. The platelets in your blood caused the bleeding to stop. **Platelets** are blood cells that clump together when they are exposed to the air. They react with substances in the blood to produce long, sticky threads that trap blood cells. These threads form a web of blood cells. This web of cells hardens to form a *clot*. (See Figure 10.4-1.) The clot plugs up the wound and prevents further loss of blood. When a clot forms on the surface of the skin, we call it a "scab." If your platelets do not function properly, even

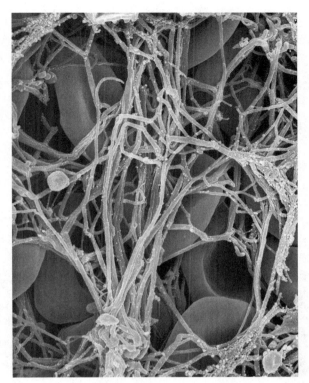

Figure 10.4-1. Formation of a blood clot.

a small cut could cause you to bleed to death.

For more information about blood cells, visit *http://www.idahoptv.org/dialogue4kids/ season4/blood/facts.html* on the Internet.

Plasma

If you remove the white blood cells, red blood cells, and platelets from a sample of blood, what remains is a straw-colored liquid called **plasma**. Plasma consists mostly of water, and it makes up more than 50 percent of blood. Dissolved nutrients, wastes, and hormones are carried by the plasma.

The Heart

For blood to do its job, it needs to *circulate*, or move around the body. The *heart* is a muscle that contracts regularly to pump or push blood. (See Figure 10.4-2.) First, blood is pumped from the heart to the lungs, where it receives oxygen and gets rid of carbon dioxide. Blood then returns to the heart, and is pumped to the rest of the body, as shown in Figure 10.4-3.

Notice that the heart consists of four separate sections. These are called chambers. The upper chambers, or **atria**, receive blood and feed it into the lower chambers, or **ventricles**. The ventricles have thick, muscular walls that pump blood. Special valves are located between the atria and the ventricles. These valves open to permit blood to flow into the ventricles. They close when the ventricles contract, so that blood does not flow back into the atria. Have you ever listened to a heartbeat? Part of that "lub-dub" sound that you hear is the sound of the valves in the heart closing, as the ventricles contract to pump blood.

Blood Vessels

Blood flows through a network of tubes called *blood vessels.* There are three types of blood vessels: arteries, veins, and capillaries. **Arteries** carry blood away from the heart, while **veins** return blood to the heart. Arteries have thick, muscular walls to prevent them from bursting as the heart pushes blood through them. Veins are not as thick as arteries, and have one-way valves to keep blood flowing in the right direction.

Connecting arteries to veins are extremely thin blood vessels called **capillaries**. It is through the capillaries that materials are exchanged between the blood and the body's cells. Dissolved nutrients,

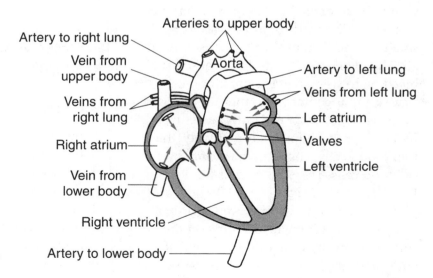

Figure 10.4-2. A human heart.

Figure 10.4-3. Pathways of blood through the circulatory system.

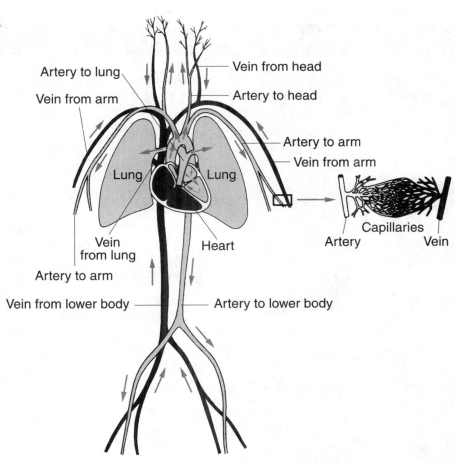

Artery to lung

Vein from arm

Vein from head

Artery to head

Artery to arm

Vein from arm

Lung

Lung

Capillaries

Artery

Vein

Vein from lung

Heart

Artery to arm

Vein from lower body

Artery to lower body

water, and oxygen pass from the blood into the cells, and some wastes (such as carbon dioxide) pass from the cells into the blood.

Lymph

Did you ever scrape your skin and notice a clear liquid in the wound? Perhaps you once developed a blister, and noticed that there was fluid directly beneath it. This fluid is called lymph. **Lymph** forms from blood plasma that filters out of capillaries and surrounds body cells. Lymph acts as a "middle man" between the blood and the body cells. After exchanging materials with body cells, lymph returns to the bloodstream through the lymph vessels. Special white cells in the lymph can destroy bacteria before they infect the bloodstream. You often notice lymph around a wound, where it guards the body from infection.

SKILL EXERCISE—*Interpreting a Diagram*

The diagram below is a schematic representation of the circulatory system. In other words, it is not a realistic drawing of body parts, but only a basic scheme of the system—the relationships among its parts and the sequence of events that occur in the system.

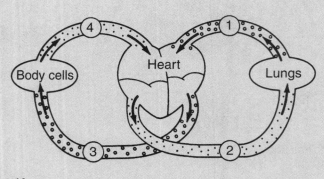

Key:

Oxygen-rich blood

Carbon dioxide-rich blood

Blood circulation is vital to the process of respiration, since blood carries oxygen to body cells and returns carbon dioxide back to the lungs to be removed.

As you have learned, arteries are blood vessels that carry blood away from the heart. Which blood vessels in the diagram are arteries? The arrows indicate that blood vessels 2 and 3 carry blood away from the heart, so they are arteries. Blood vessels 1 and 4, which return blood to the heart, are veins. Study the diagram and then answer the following questions:

1. Blood rich in oxygen is found in blood vessels

    (1) 1 and 2   (3) 1 and 3

    (2) 2 and 3   (4) 2 and 4

2. Compared with blood vessel 1, the amount of carbon dioxide in blood vessel 2 is

    (1) more   (2) less   (3) the same

3. Which statement is true?

(1) All arteries carry oxygen-rich blood.

(2) All veins carry oxygen-rich blood.

(3) Arteries from the heart to the lungs carry oxygen-rich blood.

(4) Veins from the lungs to the heart carry oxygen-rich blood.

Interesting Facts About Heart Attacks

Each year, more than 1,000,000 (one million) people have heart attacks in the United States. About 460,000 of those heart attacks are fatal. Half of those deaths occur within 1 hour of the start of symptoms and before the victim reaches the hospital. Most people wait about 2 hours after the first symptoms appear before they go a hospital emergency room. Heart attacks are still the number one cause of death in the United States.

As you already know, the heart is a muscle that pumps blood to all parts of the body. Like every other organ, the heart needs a constant supply of blood. The artery that supplies blood to the heart is called the *coronary artery. Coronary artery disease (CAD)* occurs when part of the coronary artery gets blocked. CAD results in reduced blood flow to part of the heart. This reduced blood flow damages the heart muscle, reducing its ability to pump blood. When this happens, there is generally pain in the chest area, and if the blockage is severe, there is loss of consciousness and sometimes death. Any blockage that damages the heart muscle is called a *heart attack.* A doctor using a test called an *electrocardiogram* (ee-LEK-troh-KAHR-dee-oh-grahm) can detect this damage. In some cases, people have had mild heart attacks without ever realizing that anything was wrong.

There are several factors that make heart attacks more likely. People who smoke, are overweight, have poor eating habits, and don't exercise have a greater risk of having a heart attack. Developing good habits now may prevent a heart attack in the future. To learn about the causes of heart attacks, watch the video at *http://www.health.uab.edu/default. aspx?pid=41581&site=782&return=3845*

Then visit the National Institutes of Health Web site at *http://www.nhlbi.nih.gov/health/dci/Diseases/HeartAttack/ HeartAttack_WhatIs.html* to learn bout the signs and symptoms of a heart attack. Knowing the symptoms and getting help early can save your life or the life of a relative or friend.

Questions

1. The function of the circulatory system is to
   (1) carry materials to and from the cells
   (2) break down food into a usable form
   (3) regulate body activities
   (4) respond to stimuli

2. Which group lists structures that all belong to the circulatory system?
   (1) heart, liver, and lungs
   (2) arteries, veins, and capillaries
   (3) arteries, kidneys, and stomach
   (4) skull, ribs, and muscles

3. Which blood cell fights infection?
   (1) red blood cell
   (2) white blood cell
   (3) blue blood cell
   (4) platelet

4. As oxygen from air sacs in the lungs moves into the blood, carbon dioxide from the blood moves into the air sacs. What two systems are involved in this process?
   (1) digestive system and circulatory system
   (2) respiratory system and nervous system
   (3) skeletal system and muscular system
   (4) respiratory system and circulatory system

5. Which life function is the direct responsibility of the circulatory system?
   (1) excretion   (3) transport
   (2) nutrition   (4) reproduction

6. Which figure below represents the human circulatory system?

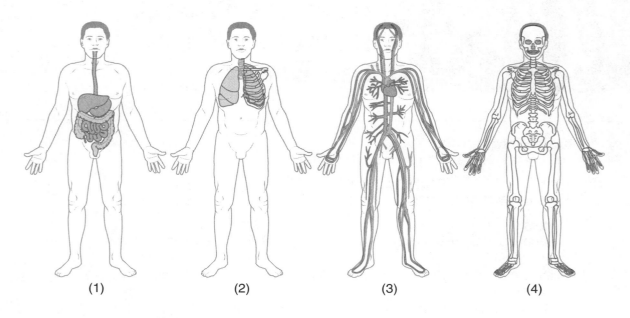

(1)   (2)   (3)   (4)

Thinking and Analyzing

1. The heart is a very important part of the circulatory system. What does the heart do?

2. Give three differences between arteries and veins.

3. The two human body systems shown below interact to perform several functions for the whole organism. Describe how gas exchange occurs when the circulatory and respiratory systems work together.

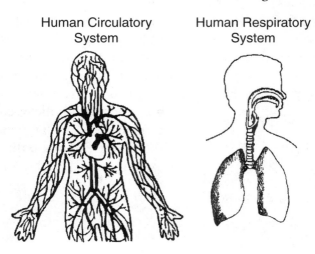

Human Circulatory System   Human Respiratory System

10.5 How Are Wastes Removed From the Body?

Objectives

Identify the organs of the excretory system.

Explain the role of the kidneys in removing wastes.

Compare excretion with elimination.

Terms

metabolism (muh-TAB-uh-liz-uhm): the sum of all chemical reactions that occur in an organism

metabolic (met-uh-BOL-ik) **wastes:** harmful products produced by chemical reactions in cells

excretion: the process of removing metabolic wastes from the body

perspiration: the fluid produced by sweat glands in the skin

feces (FEE-seez): solid wastes

elimination: the removal of solid wastes from the body

Looking Back

Let's quickly review what you learned in the last three sections. The digestive system breaks down food and supplies nutrients to the blood. The respiratory system takes in oxygen and transfers it into the blood. The circulatory system transports nutrients and oxygen to cells. Inside the cells, oxygen and sugar react to provide energy:

$$C_6H_{12}O_6 + 6O_2 \rightarrow 6CO_2 + 6H_2O + energy$$

Stop and Think: What happens to the carbon dioxide produced in cells?

Answer: The carbon dioxide is removed from cells, and eventually from the body.

Metabolism

Respiration is just one of many chemical reactions that occur inside cells. Carbon dioxide, a product of respiration, is just one of many waste materials produced inside cells. The sum of all chemical reactions that occur in the body is called **metabolism**. Many of these chemical reactions form products that can harm cells. These products are called **metabolic wastes**. **Excretion** is the process of removing metabolic wastes from the body. The system that carries out excretion is called the *excretory system.*

The Excretory System

The excretory system contains many different organs. Among the most important are the *lungs, skin, liver,* and *kidneys.*

1. The **lungs** remove carbon dioxide and water vapor from your body each time you exhale.

2. The **skin** expels wastes when you perspire. Sweat glands, deep in the skin, excrete **perspiration**. Perspiration, also known as *sweat,* is a liquid waste consisting mostly of water and salts. Perspiration leaves the body through *pores,* which are tiny openings in the surface of the skin. (See Figure 10.5-1.)

3. The *liver* produces *urea,* a waste that results from the breakdown of proteins. The liver also removes harmful substances from the blood.

4. The *kidneys* are part of the urinary system. (See Figure 10.5-2.) This system filters wastes from the blood and removes them from the body. The urinary system consists of the kidneys,

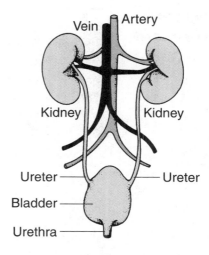

Figure 10.5-2. The urinary system.

bladder, ureters, and urethra. When blood flows through the kidneys, the body's excess water, salts, urea, and other wastes are removed from it. The kidneys function as filters, allowing some materials through, but removing others. These removed substances make up a fluid called *urine.* Urine travels through a tube called the *ureter* to the *bladder.* It is stored in the bladder until it can be excreted from the body through the urethra.

Elimination

One type of waste that we have not yet discussed in this chapter is solid waste. Solid wastes, or **feces**, contain undigested food, old red blood cells, water, and bile. Unlike the wastes removed by the excretory system, these wastes were not produced through metabolism. The process of removing solid wastes is called **elimination**, or *egestion,* and is a function of the digestive system. Elimination is *not* excretion.

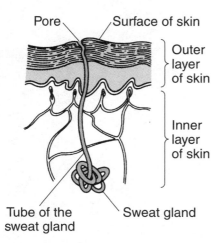

Figure 10.5-1. Sweat glands remove waste through pores in the skin.

Interesting Facts About Skin

Did you know that the skin is the largest organ in the human body? It is the container that holds your body together. It is also, as you have just learned, an important part of the excretory system, removing water and salts from the body. Did you also know that it is your skin that maintains your body temperature? It does this by producing perspiration, which cools the body when it evaporates. Skin is an organ that has many important functions!

Questions

1. Which organ is part of both the excretory system and the digestive system?
   (1) liver
   (2) lungs
   (3) skin
   (4) kidneys

2. The excretory system includes the
   (1) kidneys, liver, lungs
   (2) lungs, trachea, nose
   (3) stomach, esophagus, liver
   (4) heart, arteries, veins

3. Which of the following helps remove wastes from the body?
   (1) heart
   (2) skin
   (3) platelets
   (4) stomach

4. The removal of solid waste by the digestive system is called
   (1) excretion
   (2) filtration
   (3) respiration
   (4) elimination

5. Solid materials that are not digestible are eliminated from the body as
   (1) urine
   (2) perspiration
   (3) lymph
   (4) feces

Thinking and Analyzing

Use the diagram below to answer questions 1 through 3.

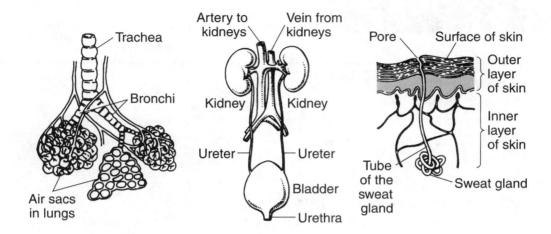

1. To which system do all three structures belong?

2. Which organ filters the blood and removes urea?

3. What waste product is removed by all of these structures?

4. Water, carbon dioxide, and urea are all metabolic wastes. Match each of these waste products to its path through the body, using the choices below:
   A. cell → blood → lung
   B. liver → blood → kidney
   C. cell → blood → skin

5. Complete the concept map at the right by naming the excretory organs represented by A, B, C, and D.

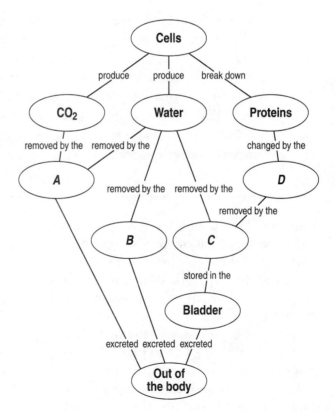

10.6 How Do We Move?

Objectives

Identify the different types of connective tissue.

Describe the role of muscles, tendons, and ligaments in moving the body.

Identify three different types of muscle.

Explain the functions of the skeletal and muscular systems.

Terms

connective tissue: tissue that supports the body and holds it together

bone: hard, rigid connective tissue that supports and protects the organs of the body

cartilage (KAR-tuh-lij): flexible connective tissue that cushions bones

ligament: connective tissue that connects bones to other bones

tendon: cord-like tissue that connects muscles to bones

muscles: masses of tissue that contract to move bones and organs

voluntary muscles: muscles that are controlled by your conscious mind

involuntary muscles: muscles that are *not* controlled by your conscious mind

cardiac muscle: involuntary muscle found only in the heart

smooth muscle: involuntary muscles involved in digestion, respiration, and other body functions

You have learned about digestion, respiration, circulation, and excretion, life functions that are common to all living things. *Stop and Think:* What life function is necessary for human survival, but not for the survival of trees?

Answer: Locomotion—the movement of an organism from place to place. *Why do you need to move?* There are many reasons: to find food, to find shelter, and to escape danger. *How do you move?* The skeletal, muscular, and nervous systems are the systems that help you to move from place to place.

The Skeletal System

The human *skeletal system*, shown in Figure 10.6-1, supports and protects the body and its organs. The skeletal system includes the *skull, spinal column, breastbone, ribs,* the *bones* of the *limbs* (arms and legs), and *cartilage.*

Figure 10.6-1. The human skeletal system.

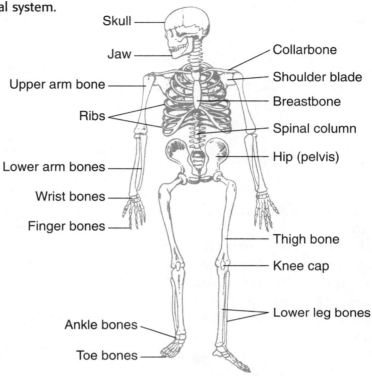

Skull

Jaw

Upper arm bone

Ribs

Lower arm bones

Wrist bones

Finger bones

Ankle bones

Toe bones

Collarbone

Shoulder blade

Breastbone

Spinal column

Hip (pelvis)

Thigh bone

Knee cap

Lower leg bones

Bones and Cartilage

Bones and cartilage are types of **connective tissue**. Connective tissues support the body and hold it together. The strongest type of connective tissue is bone tissue. **Bones** are hard and rigid. They support and protect the organs of the body. The skull protects the brain, the rib cage protects the lungs and heart, and the backbones, or spinal column, protects the spinal cord.

    Cartilage is a softer, more flexible tissue than bone. Cartilage acts as a cushion between bones, and provides flexibility at the ends of bones. Your ears and nose are flexible, because they are made of cartilage, and not bone. Rings of cartilage surround the trachea (windpipe) to keep it from collapsing, so that oxygen can always flow freely through the respiratory system.

Joints

Where one bone meets another bone, a *joint* is formed. Most joints, such as the knee and elbow, allow bones to move. However, some joints, like those in the skull, do not allow movement. Figure 10.6-2 on page 312 shows three types of joints.

Ligaments and Tendons

A **ligament** is a tough, flexible, connective tissue that connects bones to other bones. A **tendon** is a cord-like tissue that connects bones to muscles. When skeletal muscles

Figure 10.6-2. Three types of joints.

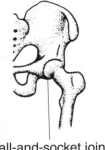

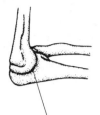

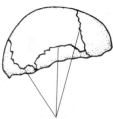

Ball-and-socket joint
(at hip)

Hinge joint
(at elbow)

Immovable joints
(in skull)

contract, tendons pull on bones, causing them to move.

Many common sports injuries involve connective tissue. "Tennis elbow" is an inflammation of the tendons around the elbow. An inflammation of any tendon is called *tendonitis.* A more serious injury occurs when a tendon is torn. One such injury is a torn Achilles tendon. This tendon is located in the back of the lower leg, where it connects to the bones of the foot. A torn tendon must be reconnected by a surgeon. One of the most common football injuries is a torn ligament in the knee. Although a torn ligament may not require surgery, it can take several months to heal.

The Muscular System

What really happens when you "make" a muscle? The muscle actually contracts—it gets shorter, thicker, and harder. **Muscles** are masses of tissue that contract to move bones or other parts of your body. Muscles that move bone are called skeletal muscles. The biceps muscle shown in Figure 10.6-3 is a skeletal muscle that bends the arm at the elbow.

The *muscular system* consists of two main kinds of muscles: voluntary and involuntary. Skeletal muscles are **voluntary muscles,** muscles that are controlled by your conscious mind. These muscles work together with the skeleton to move body parts. (See Figure 10.6-4.) An old saying goes, "It takes 37 muscles to frown, and 22 muscles to smile. So smile; it conserves energy." While it is more fun to smile than frown, the number of muscles to do either is actually about the same. The muscles in your face and around your eyes needed to smile or frown are also voluntary muscles.

Stop and Think: As you read this sentence, are you thinking to yourself,

Figure 10.6-3. Showing off biceps.

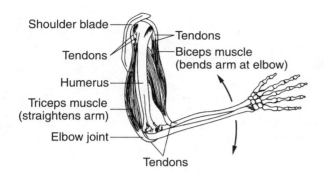

Figure 10.6-4. Muscles, tendons, and bones of the upper arm.

Table 10.6-1. The Three Types of Muscles

Type of Muscle	Striated (striped)	Voluntary
Skeletal	Yes	Yes
Cardiac	Yes	No
Smooth	No	No

"Breathe in, breathe out?" Are you consciously telling your heart to keep beating? **Answer:** Of course not. Yet your heart keeps beating, and you keep on breathing.

The beating of your heart and your breathing are automatic, involuntary acts that are controlled by involuntary muscles. **Involuntary muscles** are not controlled by your conscious mind. They work whether you want them to or not. There are two types of involuntary muscle: cardiac and smooth. The heart is made of **cardiac muscle**. Cardiac muscle is a strong, involuntary muscle found *only* in the heart. **Smooth muscle** is found in the respiratory, circulatory, and digestive systems. These muscles aid in breathing, controlling blood flow, and *peristalsis*, the movement of food through the digestive tract. The diaphragm is

one example of a smooth, involuntary muscle. Under a microscope, both voluntary muscle and cardiac muscle tissues have stripes called *striations* (stry-AY-shunz). Smooth muscle does not have these stripes, and that is why it is called "smooth." Table 10.6-1 compares the three types of muscles.

Who's the Boss?

A voluntary muscle contracts when you want it to. How does your message get to the muscle? And who is telling the involuntary muscle when to contract or relax? The "boss" of all movement in the body is the nervous system. Voluntary movements are controlled by the brain, which sends messages through the nerves to the muscles. Involuntary movement is controlled by another section of the brain, or by the spinal cord. In the next lesson, you will learn more about the nervous system.

Activity

Your fingers move because muscles in your arm pull on tendons. The tendons connect muscles to the bones in your fingers. You can see for yourself how tendons work. Place your arm flat on a table with your palm facing up. Relax your fingers. Now use your other hand to press on your arm, at a point about one-third of the way up from the wrist to the elbow. What happens to your fingers? The tendons in your arm are connected to the bones in your fingers. Pressing on these tendons pulls on these bones, causing them to move.

Interesting Facts About the Skeletal System

Did you know that there are 206 bones in the human body? The largest of these is the femur (FEE-muhr), or thighbone. The smallest is the stirrup bone, found inside the ear. Each hand contains 26 bones. Your bones continue to grow until you reach your full height. For boys, this generally occurs between ages 16 and 20. Girls generally reach their full height somewhat earlier.

Have you ever seen a human skull? It is missing a nose and both ears! Why? The material in bones can last long after the death of the individual. Cartilage is not as durable. The nose and outer ears contain no bones, just cartilage. Make no bones about it, your nose and ears will disappear soon after you are gone!

Questions

1. Which two systems of a rabbit's body must be working together for the rabbit to run away from a fox?
   (1) digestive and endocrine systems
   (2) reproductive and nervous systems
   (3) muscular and skeletal systems
   (4) excretory and respiratory systems

2. Which body system supports and protects other body systems?
   (1) skeletal   (3) reproductive
   (2) endocrine   (4) digestive

3. Which body system provides movement for the body?
   (1) digestive   (3) muscular
   (2) circulatory   (4) endocrine

4. Which group lists three parts of the skeletal system?
   (1) heart, stomach, brain
   (2) tendons, nerves, brain
   (3) bones, nerves, blood
   (4) cartilage, ligaments, bones

5. Which type of muscle is found only in the heart?
   (1) voluntary   (3) cardiac
   (2) smooth   (4) involuntary

6. Which activity is most likely to be controlled by a smooth muscle?
   (1) breathing   (3) chewing
   (2) walking   (4) thinking

7. The diagram below best demonstrates that
   (1) the skeleton protects body organs
   (2) bones are held together at joints by ligaments
   (3) muscles and bones work together to move body parts
   (4) cartilage protects and cushions bones

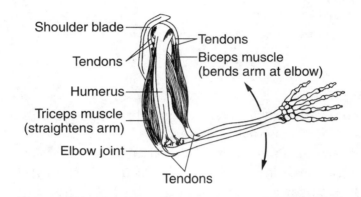

Thinking and Analyzing

1. Which three systems of the human body function together to move and control body parts?

2. Describe one similarity and one difference between tendons and ligaments.

3. What are three functions of the skeletal system?

4. Complete this paragraph, using the terms "smooth," "cardiac," and "voluntary."

   As the _____ muscles of his stomach churned and digested food while he was returning home from his favorite restaurant, Paul noticed a pack of mean-looking dogs following him. The _____ muscles in his heart pumped faster. He realized that the _____ muscles in the dogs' legs could move the dogs very quickly, so he knew not to run away. As the dogs moved closer, the _____ muscle called the diaphragm caused Paul's breathing to speed up. Fortunately, the dogs passed by without even noticing him.

10.7 How Do We Respond to Change?

Objectives

Identify the parts of the nervous system.
Describe the role of endocrine glands.
Describe how reflexes work.

Terms

reflex (RHEE-flex): an automatic response designed to protect the body
central nervous system: the brain and spinal cord
neurons (NUH-rahnz): nerve cells
sense organs: organs that receive information about the environment
synapse (SIH-naps): the gap between one neuron and the next
stimulus: an environmental change that causes a response
endocrine (EHN-doh-krin) **gland**: an organ that secretes hormones
hormone (HAWR-mone): a substance secreted by an endocrine gland that has an effect on the body

---

Movement is one way your body can respond to changes in your environment. You can run from danger, move into the shade, or get up and put on a warm sweater. When you move, your nervous system sends the necessary messages to your muscles. Movement, however, is not the only way your body reacts to changes in its environment.

Regulation

The coordination of the various organ systems is called *regulation*. Regulation allows organisms to respond to changes both inside and outside their bodies,

maintaining homeostasis. Regulation is controlled by two related systems: the nervous system and the endocrine system. Both systems can detect changes. They send messages to the parts of the body that can respond to those changes. The nervous system sends electrical and chemical messages through the nerves. The endocrine system sends chemical messages through the blood.

The Nervous System

The nervous system consists of the brain, spinal cord, nerves, and parts of the sense organs. (See Figure 10.7-1.)

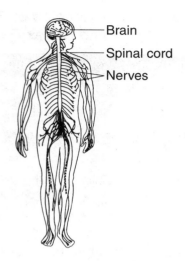

Figure 10.7-1. The nervous system.

The Central Nervous System

Together, the brain and the spinal cord make up the **central nervous system**. The central nervous system controls all body activities.

The *brain* receives and interprets *nerve impulses* ("messages"). It controls thinking, voluntary action, and some involuntary actions, such as breathing and digestion. For more information about your brain, go to the Web site *http://www.idahoptv.org/ dialogue4kids/season3/brain/facts.html*

The *spinal cord* channels nerve impulses to and from the brain, and controls many automatic responses, or *reflexes,* such as the pulling away of your hand from a hot flame.

Sensing and Responding

Special cells called **neurons**, or nerve cells, receive and deliver messages. (See Figure 10.7-2.) Nerve cells provide a means of communication between the brain and

spinal cord and the body's sense organs, muscles, and glands. The **sense organs** include the skin, eyes, ears, nose, and tongue. This communication is accomplished through electrical and chemical messages. Electrical messages travel along the neuron until they reach another neuron, a muscle, or a gland. There is a gap called a **synapse** between one neuron and the next. The electrical message cannot travel across the gap. Chemicals are released by one neuron to transmit the message across the synapse to the next neuron.

Neurons transmit messages to and from the central nervous system so that the body can respond to a **stimulus**. A stimulus is any change in the environment that causes a response. *Sensory neurons* are nerve cells that carry information about a stimulus *to* the central nervous system. *Motor neurons* are nerve cells that carry information about a response *away from* the central nervous system.

A sense organ first receives a stimulus from the environment. Sensory neurons carry that information to the brain or spinal cord. The central nervous system receives the information and decides what action to

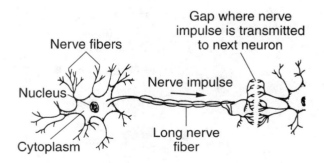

Figure 10.7-2. A neuron sends and receives messages.

take. Motor neurons then carry a message to muscles or glands, which respond to the stimulus.

Reflexes

Do you have good reflexes? Has your doctor ever tapped your knee with a small rubber hammer to test your reflexes? A **reflex** is an automatic response to a particular stimulus. Reflexes are designed to allow you to respond quickly to protect your body from harm.

For example, if you accidentally touch a hot object, you pull your hand away even before you realize that it is hot. The stimulus is the hot object, and the response is pulling your hand away. The response is called a reflex only when it is involuntary. If dust gets into your nose you might take out a tissue and blow your nose. That is a response to a stimulus, but it is not a reflex.

Table 10.7-1. Examples of Stimuli and Reflex Responses

Stimulus	Reflex Response
Tap on the kneecap	Leg jerks upward
Particle in the eye	Blinking
Dust in the nose	Sneezing
Smell of food	Salivating
Touching a hot object	Pulling away from the object

On the other hand, if the dust caused you to sneeze, that would be a reflex. Table 10.7-1 lists some stimuli and reflex responses.

The Endocrine System

The endocrine system is made up of glands. An **endocrine gland** is an organ that secretes

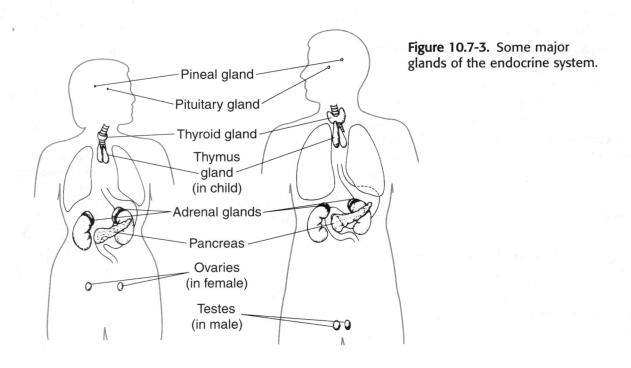

Figure 10.7-3. Some major glands of the endocrine system.

Pineal gland

Pituitary gland

Thyroid gland

Thymus gland (in child)

Adrenal glands

Pancreas

Ovaries (in female)

Testes (in male)

hormones. A hormone is a substance that has a specific effect on the body. Figure 10.7-3 shows some endocrine glands. When an endocrine gland secretes a hormone into the bloodstream, the blood carries the hormone to an organ. The organ responds to that hormone in a specific way. For example, if you are suddenly faced with danger, such as a snarling dog, the hormone *adrenaline* is released by your *adrenal gland*. The adrenaline makes your heart beat faster and your breathing more rapid. More sugar is released into your bloodstream to provide energy. These changes prepare your body to respond to the danger. Your most likely response to this danger would be to run away.

You probably are noticing changes in your body as you begin to enter your teen years. Hormones produced in the pituitary gland are causing you to grow taller. Sex hormones, secreted by the testes or the ovaries, are causing other changes in your body. This period of hormonal change in both boys and girls is called *puberty* (PYOO-buhr-tee).

Activity

Is one side of your brain more dominant than the other? Take this 50-question quiz and find out:

http://library.thinkquest.org/19910/data/brain_test

Interesting Facts About the Brain

Do you have one brain or two? You have only one brain, but it is divided into two sides, called hemispheres. Each hemisphere controls different body functions. For example, the right side of the brain controls the muscles on the left side of the body, while the left side of the brain controls the muscles on the right side of the body.

The two hemispheres of the brain also divide the work when it comes to thinking. The right side of the brain controls things like music, art, and creativity. The left side of the brain controls language and math. The right brain deals with faces, places, and objects. The left brain deals with letters, numbers, and words.

Questions

1. Hormones are chemicals that are secreted by the
   (1) gallbladder
   (2) brain
   (3) endocrine glands
   (4) small intestine

2. The endocrine system works with the nervous system to
   (1) digest nutrients
   (2) exchange gases with the environment
   (3) produce energy
   (4) regulate body activities

3. The brain, spinal cord, and sensory neurons are all part of the
   (1) nervous system
   (2) respiratory system
   (3) circulatory system
   (4) endocrine system

4. The human cell shown in the diagram below

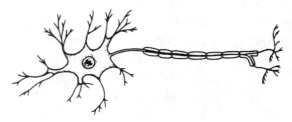

   (1) stores excess food
   (2) sends and receives nerve impulses
   (3) covers and protects the body
   (4) carries oxygen to other cells

5. A change in the environment that causes a response is known as a
   (1) stimulus   (3) reflex
   (2) habit   (4) source

6. A tissue designed to carry messages throughout the body is most likely to be
   (1) skin   (3) nerve
   (2) muscle   (4) bone

Thinking and Analyzing

1. Compare the way messages are carried in the nervous system to the way messages are carried in the endocrine system.

2. When you touch a baby's cheek, it turns its head toward you. This automatic response is an example of a reflex. What is the stimulus and what is the response for this reflex?

Review Questions

Term Identification

Each question below shows two terms from Chapter 10. One of the terms is defined.

(1) Choose the term that matches the definition.

(2) Describe how the two terms are different.

Following each term is the section (in parenthesis) where the description or definition of that term is found.

1. *Mechanical digestion (10.2) — Chemical digestion (10.2)*
   The physical breakdown of food into smaller pieces

2. *Nutrients (10.2) — Digestive enzymes (10.2)*
   Substances that aid in chemical digestion

3. *Inhale (10.3) — Exhale (10.3)*
   Breathe out

4. *Alveoli (10.3) — Diaphragm (10.3)*
   A muscle in the chest that controls breathing

5. *Platelet (10.4) — Plasma (10.4)*
   A blood cell that causes blood to clot

6. *Atria (10.4) — Ventricles (10.4)*
   The upper chambers of the heart

7. *Arteries (10.4) — Veins (10.4)*
   Blood vessels that carry blood away from the heart

8. *Excretion (10.5) — Elimination (10.5)*
   Removal of solid wastes from the body

9. *Bones (10.6) — Cartilage (10.6)*
   Hard rigid tissue that supports and protects the organs of the body

10. *Ligaments (10.6) — Tendons (10.6)*
    Cord-like tissue that connects bones to muscles

11. *Cardiac muscle (10.6) — Smooth muscle (10.6)*
    Involuntary muscle that is found in many organ systems, including the digestive and respiratory systems

12. *Stimulus (10.7) — Reflex (10.7)*
    A change that causes a response

13. *Neuron (10.7) — Hormone (10.7)*
    A nerve cell

Multiple Choice (Part 1)

Questions 1 and 2 refer to the four diagrams of organ systems shown below.

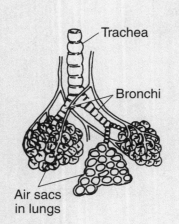 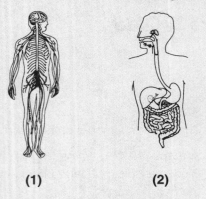 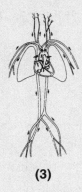 

(1)   (2)   (3)   (4)

1. Which diagram represents the digestive system?
(1) 1   (3) 3
(2) 2   (4) 4

2. Which system transports needed nutrients to all the cells of the body?
(1) 1   (3) 3
(2) 2   (4) 4

3. The process that releases energy from nutrients is called
(1) cellular respiration (3) digestion
(2) excretion   (4) circulation

4. Cellular respiration takes place in
(1) the blood only   (3) the heart only
(2) the lungs only   (4) all body cells

5. Which organ belongs to both the excretory system and the respiratory system?
(1) heart   (3) lungs
(2) kidney   (4) liver

6. The excretory system includes the
(1) kidneys, liver, lungs
(2) lungs, trachea, nose
(3) stomach, esophagus, liver
(4) heart, arteries, veins

7. Which of the following helps remove wastes from the body?
(1) the skull   (3) the spinal cord
(2) the skin   (4) the stomach

Use the diagram below to answer question 8 on the next page.

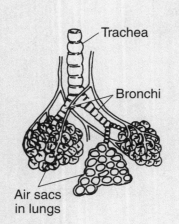 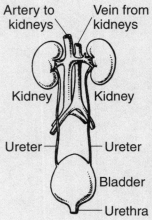 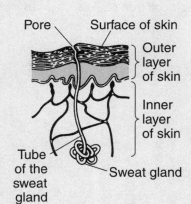

8. To which system do all three structures belong?
(1) excretory   (3) respiratory
(2) circulatory   (4) skeletal

9. Once you have taken a bite of an apple, which of the following represents the correct pathway of its nutrients?
(1) circulatory system → cell → digestive system
(2) cell → digestive system → circulatory system
(3) digestive system → circulatory system → cell
(4) circulatory system → digestive system → cell

10. The endocrine system produces substances that affect organ functions. These substances are called
(1) nutrients   (3) microbes
(2) hormones   (4) wastes

11. Nutrients from digested food enter the bloodstream through the process of
(1) absorption   (3) respiration
(2) elimination   (4) secretion

12. The kidneys, which remove dissolved wastes from the blood, are organs of the
(1) endocrine system
(2) excretory system
(3) skeletal system
(4) nervous system

13. Which three systems of the human body function together to move and control body parts?
(1) nervous, skeletal, and muscular
(2) muscular, endocrine, and excretory
(3) digestive, excretory, and reproductive
(4) circulatory, endocrine, and respiratory

14. The labeled organs in the diagram below are part of which human body system?
(1) respiratory   (3) digestive
(2) endocrine   (4) circulatory

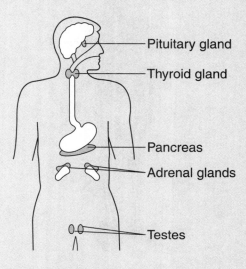

Pituitary gland

Thyroid gland

Pancreas

Adrenal glands

Testes

15. Which human body systems are directly involved in reflex actions, such as knee jerk, blinking, and jumping when startled?
(1) circulatory and respiratory
(2) digestive and excretory
(3) nervous and muscular
(4) reproductive and skeletal

16. Which body system is responsible for the elimination of liquid and gaseous wastes?
(1) nervous   (3) excretory
(2) skeletal   (4) digestive

17. The diagram below shows a human body system.

What are two functions performed by this body system?
(1) protect and support the body
(2) produce and transport oxygen within the body
(3) produce and excrete waste products
(4) control and coordinate body activities

Thinking and Analyzing (Part 2)

1. What is the main function of the human circulatory system?

2. The diagram at the right represents a magnified view of an air sac in the human lung. The white arrows indicate blood flow. What two body systems are interacting in this diagram?

Base your answers to questions 3 through 5 on the passage below and on your knowledge of science.

The walls of the small intestines are covered with finger-like projections called *villi*. These villi increase the area of contact between digested food and capillaries.

3. What is the name of the process that occurs at the villi?

4. Which body system contains villi?

5. What structures in the lungs perform a similar task to that of villi?

6. The table below shows the symptoms of some diseases caused by bacteria. Which two diseases affect the digestive system?

Disease	Symptoms
botulism	vomiting, abdominal pain, coughing, muscular weakness, visual disturbance
pneumonia	inflammation of lungs, fever, shortness of breath, fluid in lungs
typhoid fever	red rashes, high fever, intestinal bleeding
tetanus	uncontrolled contractions of voluntary muscles

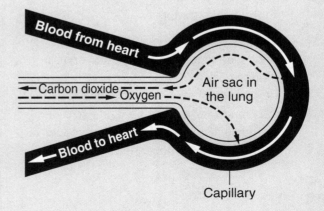

7. Use the Venn diagram at the right to organize the list of organs into the systems to which they belong. If an organ is part of two systems, place it in the area where the two systems overlap.

Kidneys, Urethra, Stomach, Esophagus, Liver, Lungs, Skin, Small Intestine, Gallbladder, Nose, Trachea

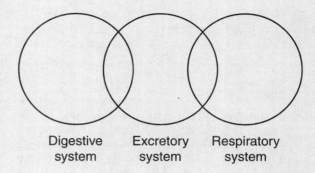

Digestive system Excretory system Respiratory system

Chapter Puzzle (*Hint:* The words in this puzzle are terms used in the chapter.)

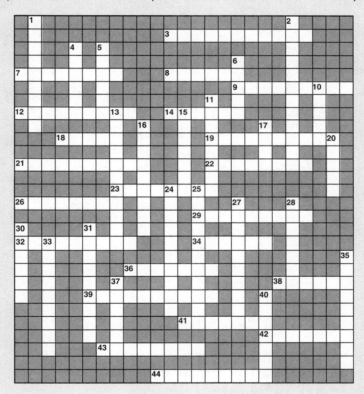

Across

3 the tiniest blood vessels that connect veins and arteries

7 the reaction that produces energy in the mitochondria is called ___ respiration

8 breathe out

9 connective tissue that surrounds and cushions bones

12 muscles that move when you want them to (e.g., your arm and leg muscles) are called ___ muscles

14 products of metabolism that must be removed from the body

18 blood vessels that carry blood to the heart

19 ___ digestion involves breaking down food into smaller pieces

21 ligaments, bones, tendons, and cartilage are examples of ___ tissue

22 connects one bone to another

23 blood vessels that carry blood away from the heart

26 the type of blood cell that stops bleeding by causing clotting

29 an illness or sore caused by bacteria, viruses, or other tiny organisms is called an ___

32 hormones are secreted by the ___ glands

34 breathe in

36 the brain and the spinal cord together make up the ___ nervous system

38 a type of connective tissue that connects muscle to bone

39 a chemical found in red blood cells that carries oxygen

41 digestive ___ are chemicals that aid in chemical digestion

42 the muscle that is found only in the heart

43 the form of digestion in which substances in food are broken down to simpler substances that can be used by the body

44 useful chemicals found in food

Down

1 the process of producing chemicals and releasing them into the body

2 a liquid produced by the skin

4 nerve cell

5 the liquid part of the blood

6 solid wastes

10 the upper chambers of the heart

11 a change in the environment that causes a response

13 the process of taking in oxygen, using it to provide energy, and removing carbon dioxide

15 tiny air sacs in the lungs

16 the process of breaking food down to a form that can be used by the body

17 animals that maintain a constant body temperature are called ___-blooded animals

20 a clear fluid that bathes the cells

24 the coordination of the various organ systems

25 removal of solid wastes

27 the sum of all of the chemical reactions that occur in the body

28 adrenaline is an example of a ___, a chemical that has a specific effect on the body

30 an automatic response to a stimulus

31 moving air in and out of the lungs

33 a muscle that controls breathing

35 the lower chamber of the heart that pumps blood into the arteries

37 the type of involuntary muscle found in the stomach or diaphragm

40 masses of tissue that contract to move bones

Unit 4

Dynamic Equilibrium: Other Organisms

Part 1—Essential Question

This unit focuses on the following essential question:

How is homeostasis maintained in other organisms?

You have just learned about homeostasis in one organism, humans. You learned how human systems carry out life functions. In this unit, we will be examining how other organisms maintain homeostasis. You will discover that while the systems may vary, the functions stay the same.

This unit is organized by kingdom. The first chapter in the unit begins with the systems found in our closest relatives—the animals. In the second

chapter, we discuss how plants maintain homeostasis. The last chapter discusses the tiniest of all organisms, those containing just a single cell. As you examine these tiny organisms, you will discover that they occupy three different kingdoms. The scientific names of the five kingdoms are: *animalia, plantae, fungi, protista,* and *monera.*

As you go through these three chapters, and look at the many kinds of living things, you should notice how very different they are in appearance and structure. Yet you will find that they are very similar in what they do to stay alive. The great variety of living things, called biodiversity, makes biology the fascinating science that it is.

Part II—Chapter Overview

Chapter 11 discusses how animals maintain homeostasis. You have already learned about one animal, humans. In Chapter 11, you will see how other animals, large and small, carry out life functions. The focus in the chapter is on those animals that are very different from humans. You will learn about insects, earthworms, fish, and frogs. Each lesson focuses on one or two organ systems, and demonstrates how different animals maintain homeostasis.

Chapter 12 discusses how plants maintain homeostasis. Unlike animals, plants do not feed on other organisms; they make their own food. You will look at the special structures that allow plants to use sunlight, oxygen, and water to produce sugar. You will see how leaves, stems, and roots work together to transport materials throughout the plant. You will also see how plants can respond to changes in their environment by growing in a particular direction.

Chapter 13 looks at one-celled, or unicellular, organisms. You will see how single-celled organisms carry out their life functions. One of these life functions is locomotion; one-celled organisms move around in a variety of different ways. These tiny organisms also have special structures that enable them to eat and excrete.

When you complete Unit 4, you will have explored the great diversity of living things on Earth.

Chapter 11

Other Animals

Contents

A monarch butterfly is just one of more than a million different species of animals.

What Is This Chapter About?

In Chapter 10, you learned how human organ systems perform life functions to maintain homeostasis. There are millions of other species of animals, and they all must maintain homeostasis as well. All organisms must carry out the same life functions that humans do. This chapter takes a look at just a few different kinds of animals and compares how they perform life functions. You will learn how different animals use different structures to accomplish the same goals.

In this chapter you will learn:

1. All animals must carry out the same life processes to maintain homeostasis.

2. Animals may have an internal skeleton, an external skeleton, or no skeleton.

3. Animals may take in oxygen using lungs, gills, spiracles, or moist skin.

4. Nutrition and excretion are accomplished in many different ways.

5. Animals respond to changes in their environment.

Science in Everyday Life

What is the most unusual pet you have ever seen? People keep a variety of animals. Most pet owners have dogs or cats or both. Others have unusual pets, such as snakes, turtles, frogs, tarantulas, snails, mice, angelfish, lizards, parrots, sugar gliders, rabbits, hamsters, clownfish, and guinea pigs. As a pet owner, you need to find the correct food, provide the proper temperature, and in many cases, remove wastes from the animal's environment. No matter what kind of pet you choose, understanding its life functions will make you a better pet owner.

Internet Sites:

http://www.kidport.com/reflib/science/animals/animals.htm This Web site explores the animal kingdom. It has many links to other interesting Web sites. Start by comparing vertebrates and invertebrates. Follow the links to explore the diversity of living things.

http://www.globio.org Learn about biodiversity by selecting an area of the world or a type of animal. This site works better if you have Flash 8 player installed on your computer.

11.1

How Are Animals Classified?

Objectives

Distinguish between vertebrates and invertebrates.

Explain biodiversity.

Terms

mammal: a warm-blooded animal that has hair covering its body and provides milk for its young

vertebrae (VER-tuh-bray): bones that surround and protect the spinal cord

vertebrate (VER-tuh-brit): an animal that has a backbone

invertebrate (in-VER-tuh-brit): an animal that does not have a backbone

biodiversity (buy-oh-di-VUR-si-tee): variety in living things

---

Think of three animals. Perhaps you thought of a dog, a cat, a lion, or a giraffe. All of these animals are **mammals**. Mammals, including humans, are warm-blooded animals that have hair covering their bodies and provide milk for their young. When we hear the word "animal," we often think first of mammals. There are, however, many other types of animals: birds, insects, worms, snails, lizards, and frogs, just to name a few.

Classification

All living things are classified based on their structure and how they perform life functions. Each different type of organism may be placed in one of five main categories, called kingdoms. The five kingdoms are animals, plants, fungi,

protists, and monera. Table 11.1-1 gives examples of, and describes the differences among these kingdoms. This chapter focuses on the animal kingdom.

Biodiversity

Like all kingdoms, the animal kingdom is divided into smaller groups, based on similar structures. Fish, amphibians, reptiles, birds, and mammals share similar structures called **vertebrae**—bones in their backs that surround the spinal cord. Animals with these backbones are called **vertebrates**. Animals without backbones are called **invertebrates**. Invertebrates include earthworms, lobsters, grasshoppers, spiders, jellyfish, and snails. These animals actually have no bones at all. (See Figure 11.1-1.)

Table 11.1-1. The Five Kingdoms of Living Things

Kingdom	Scientific Name	Description	Examples
Animals	Animalia	Multicellular animals that have cells containing centrioles	Insects, fish, birds, mammals, reptiles
Plants	Plantae	Multicellular plants composed of cells that have a cell wall and contain chloroplasts (to make their own food)	Trees, shrubs, grasses, mosses
Fungi	Fungi	One-celled and multicellular organisms made up of cells that have a cell wall, but no chloroplasts	Yeast, molds, mushrooms
Protists	Protista	Mostly one-celled plant-like and animal-like organisms	Algae, protozoa (e.g., ameba, paramecium)
Monera	Monera	One-celled organisms that lack an organized, membrane-enclosed nucleus	Bacteria (e.g., *E.coli*)

There are over one million different animal species on Earth. Most of these species are invertebrates. Biologists use the word **biodiversity** to describe the enormous variety among living things. However, all animals, vertebrate and invertebrate alike, have something in common. They must all carry out the same life functions. In the rest of this chapter, we will compare just a few different animals and the structures they use to maintain homeostasis.

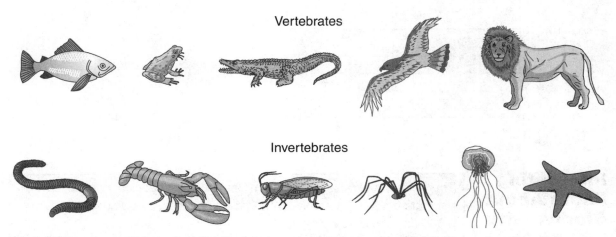

Vertebrates

Invertebrates

Figure 11.1-1. Animals can be classified as vertebrates (animals with backbones) or invertebrates (animals without backbones). Examples of vertebrates are the tuna, toad, crocodile, eagle, and lion. Examples of invertebrates are the earthworm, lobster, grasshopper, spider, jellyfish, and sea star.

SKILL EXERCISE—*Constructing a Bar Graph*

1. We used the data in the Interesting Facts About Biodiversity feature to create the bar graph below. Copy this graph onto a piece of graph paper, and complete the graph by doing the following:

a. give the graph a title

b. label the *y*-axis

c. label the *x*-axis

d. below each bar, write the name of the type of animal it represents

You may use the graph from questions 5 and 6 on page 333 at the end of this lesson as a guide.

2. Which type of animal is listed in the Interesting Facts About Biodiversity feature, but not shown in the graph?

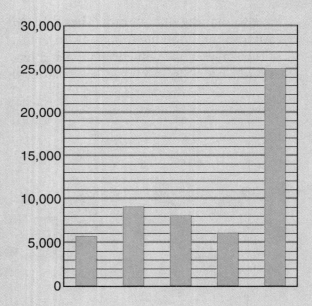

Interesting Facts About Biodiversity

There are about 1,000,000 (one million) different species of insects. By comparison, there are only 25,000 species of fish; 9,000 species of birds; 8,000 species of reptiles; 6,000 species of amphibians; and 5,500 species of mammals, including one species called humans. Scientists estimate there may be millions more insect species to be discovered.

Questions

1. Which of the following is a vertebrate?
   (1) grasshopper (3) earthworm
   (2) toad (4) spider

2. Which of the following animals has **no** backbone?
   (1) dog (3) alligator
   (2) hawk (4) bee

3. Which group contains the greatest number of different kinds of animals?
   (1) invertebrates (3) mammals
   (2) vertebrates (4) insects

4. Which term includes the other three?
   (1) animals (3) goldfish
   (2) fish (4) vertebrates

Use the graph below and your knowledge of science to answer questions 5 and 6. The graph shows the number of known species in the five kingdoms of living things.

5. Which kingdom contains the most diversity?
   (1) animalia (3) fungi
   (2) plantae (4) monera

6. Based on the graph, which kingdom contains approximately twice as many known species as the protista kingdom does?
   (1) animalia (3) fungi
   (2) plantae (4) monera

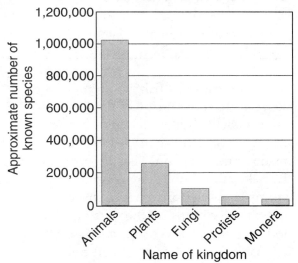

Five Kingdoms of Living Things

Thinking and Analyzing

1. Based on the graph used for questions 5 and 6, approximately how many different species of living things are found on Earth?

2. Biologists prefer the name "sea star" to "starfish" because a starfish is not a fish. Based on what you learned in this chapter, how could you prove that a sea star is not a fish?

11.2 How Do Animals Keep Their Shape?

Objectives

Distinguish between exoskeletons and endoskeletons.

Identify soft-bodied animals.

Identify the type of skeleton found in different types of animals.

Terms

endoskeleton (en-doh-SKEL-ih-tin): the system of bones that is found inside an animal

exoskeleton (ek-soh-SKEL-ih-tin): the hard outer covering found on arthropods

arthropod: an invertebrate with legs that have joints

chitin (KY-tin): the hard material that makes up the exoskeleton

When you eat most fish, you remove the skeleton from the inside of the fish. When you eat shellfish like shrimp or lobster, however, you remove the fish from the inside of the skeleton! Fish are vertebrates; their skeletons are inside. Lobsters and shrimp, however, are types of invertebrates that have skeletons on the outside of their bodies. Have you ever seen a jellyfish? Jellyfish are animals with a soft body. They have no skeletons at all. (See Figure 11.2-1.)

Endoskeleton

In Chapter 10, you learned that the human skeletal system supports and protects the body. Other vertebrates, including fish, birds, and elephants, have similar skeletal systems, as shown in Figure 11.2-2. The skeleton of vertebrate animals consists of a

hard material called bone, which maintains the shape of the animal and protects the soft internal organs. Because these bones are found *inside* the body, they are called an **endoskeleton**. (The prefix "endo" means "inside.")

Although all vertebrates have bones, not all bones are alike. Different types of

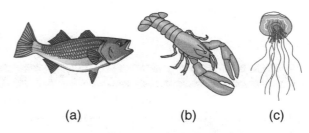

(a)   (b)   (c)

Figure 11.2-1. (a) A fish has its skeleton on the inside; (b) a lobster has its skeleton on the outside; and (c) a jellyfish has no skeleton at all.

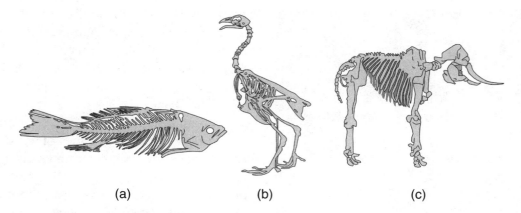

Figure 11.2-2. Endoskeletons of (a) a fish, (b) a bird, and (c) an elephant.

vertebrates have different types of bones. **Stop and Think:** How are the bones of an elephant different from the bones of an eagle? **Answer:** Large land animals, such as humans and elephants, need large, thick bones that can support their weight. The bones of most birds, however, are light and hollow so that they can fly. However, the bones of flightless birds, such as the ostrich and the penguin, are stronger and heavier than the bones of birds that can fly. Fish have light, flexible bones that allow them to swim and to stay afloat in the water.

Exoskeleton

Have you ever seen the skeleton of a lobster? You certainly have! It is shown in Figure 11.2-1. Lobsters are invertebrates because they have no bones. What they do have, however, is a hard *outer* covering called an **exoskeleton**. (The prefix "exo" means "outside.") The hard skeleton of a lobster is on the outside of its body.

Lobsters are **arthropods**, invertebrates with legs that have joints. Spiders and insects are arthropods, too. (See Figure 11.2-3.) All arthropods have exoskeletons. When you are looking at a lobster, a crab, a bee, or a beetle, what you are really seeing is its exoskeleton.

Exoskeletons are made of a material called **chitin**. Like the bones of vertebrates,

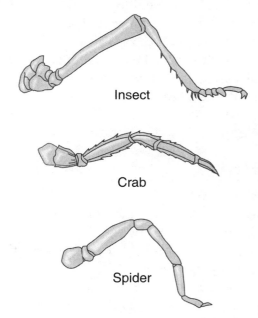

Figure 11.2-3. Arthropods have jointed legs.

Figure 11.2-4. Students viewing a butterfly collection at a museum.

chitin is a hard, durable material that maintains its appearance long after the organism dies. Since the exoskeleton completely covers the animal, the animal may appear to be unchanged many years after it dies. Have you ever seen a butterfly collection in a museum? Some butterfly collections are a hundred years old.

Stop and Think: Why do the butterflies still look the same after so many years? *Answer:* All that remains of the insect is its skeleton. Since its skeleton is *outside* its body, the insect looks the same as it did when it was alive. (See Figure 11.2-4.)

Vertebrates grow larger as their bones grow larger. Unlike the bones of vertebrates, the exoskeletons of invertebrates cannot grow. As an arthropod grows, it eventually becomes too large for its exoskeleton. The animal sheds the old exoskeleton and grows a new, larger one in a process called *molting.* A lobster may molt twenty to thirty times in its first five to seven years of life. After that, it molts about once a year.

No Skeleton

Some animals appear to have no skeletons at all. We call such animals "soft bodied." Soft-bodied animals include the jellyfish, squid, octopus, and earthworm. These animals do not have a rigid solid substance to give their bodies shape. Some soft-bodied animals are able to form hard protective shells. These include clams, oysters, and snails. These shells are not made of chitin and therefore are not considered exoskeletons.

Activity

Visit the Web site *http://www.brooklynexpedition.org/structures/bugs/insects/ins_main.html* on the Internet. Go to the "compare-a-skeleton" link under "The Amazing Structures of Insects." Then answer the following questions.

1. How far can some frogs jump?
2. Describe the skeleton of a snake.
3. Why don't fish need well-connected vertebrae?

Interesting Facts About Arthropods

There are millions of different species of arthropods, but most of these are insects. Look at the arthropods below. Can you tell which are insects and which are not?

| Spider | Firefly | Scorpion | Tick | Centipede |

In fact, only the firefly is an insect. There are many different kinds of insects, but all of them have six legs. Spiders, scorpions, and ticks have eight legs and are called *arachnids*. *Myriapods*, such as centipedes and millipedes, have at least nine pairs of legs.

Questions

1. All animals with exoskeletons and jointed legs are called
   (1) mammals   (3) vertebrates
   (2) arthropods   (4) insects

2. Complete the analogy: exoskeleton is to insect as endoskeleton is to
   (1) bone   (3) lobster
   (2) chitin   (4) bird

3. Mammals, birds, and fish are examples of vertebrates. Vertebrates have
   (1) exoskeletons   (3) backbones
   (2) chitin   (4) soft bodies

4. Both endoskeletons and exoskeletons
   (1) are made of bones
   (2) are made of chitin
   (3) are found on the inside
   (4) protect and support the internal organs

Thinking and Analyzing

1. Why do invertebrates need to molt, but vertebrates do not?

2. List two characteristics that can help you identify an arthropod.

11.3 How Do Animals Obtain Oxygen?

Objectives

Describe several methods animals use to obtain oxygen.

Explain how skin, gills, and spiracles bring oxygen to the cells.

Terms

diffusion (dih-FYOO-zhin): the movement of material from an area of high concentration to an area of low concentration

spiracles (SPIR-ih-cullz): tiny openings in the bodies of insects that take in air

gills: structures in fish that remove dissolved oxygen from water

Why Animals Need Oxygen

All organisms must obtain and use energy. Animals obtain energy through cellular respiration, the reaction of glucose with oxygen. (See Lesson 10.3.) All animals need a continuous supply of oxygen. You learned in Chapter 10 that the human respiratory system takes in air through the lungs. Oxygen passes from the lungs into the blood, and travels to all the cells in the body. All mammals, birds, and reptiles have similar respiratory systems.

Not all animals have lungs, however. In fact, not all animals have blood! Yet, in all animals, oxygen must reach every cell. In this lesson, we will explore some other ways that animals take in oxygen.

Hydra

The simplest method of getting oxygen to every cell is to place every cell in direct

contact with the source of the oxygen. A hydra is a very simple multicellular animal that lives in water. (See Figure 11.3-1.) The

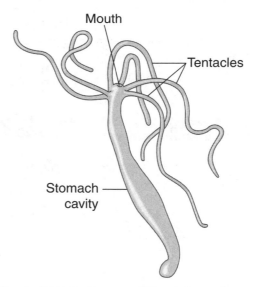

Figure 11.3-1. Oxygen diffuses from the watery environment directly into every cell of the hydra's body.

hydra's body is only two cells thick. Every cell is in contact with water. Oxygen that is dissolved in the water passes directly through the cell membranes, and into each cell.

Diffusion

The movement of material from an area of high concentration to an area of low concentration is called **diffusion**. There are more molecules of oxygen surrounding the outside of the cells (this is the area of high concentration) than there are inside the cells (the area of low concentration), so the oxygen diffuses into the inside of the cells, as shown in Figure 11.3-2. Diffusion into a cell requires a moist surface. *Stop and Think:* What

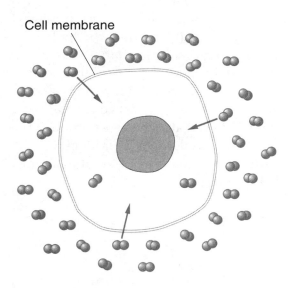

Cell membrane

 Oxygen molecule (O_2)

Figure 11.3-2. Diffusion is the movement of materials from an area of higher concentration (outside the cell) to an area of lower concentration (inside the cell).

happens to a hydra when it is removed from the water? After all, there is plenty of oxygen in the air! *Answer:* It dies because it dries up. The diffusion of oxygen into the cells must take place in a moist environment.

Earthworms

How would you describe the skin of an earthworm? Slimy? Wet? The skin of an earthworm must always be moist because the earthworm "breathes" through its skin. Oxygen diffuses from the air, through the moist skin, and into the blood. Blood carries oxygen to all of the cells of the earthworm. Why do earthworms come out of the ground after a heavy rain? The rain replaces some of the air in the soil. There is not enough oxygen in the rainwater to keep the earthworm alive. The earthworm must be careful, though. If it stays on the surface too long, its skin can dry out, and the earthworm will die.

Insects

Insects take in oxygen from the air, but their circulatory system does not bring oxygen to their cells. Insect blood does not contain hemoglobin and, therefore, cannot carry oxygen. This is also why insect blood is not red. (If you see red blood when you swat a mosquito, it is probably *your* blood.)

Insects take in air through tiny openings called **spiracles**. (See Figure 11.3-3 on page 340.) The spiracles branch out into small tubes that connect directly with all of the cells of the insect's body. The inside of these tubes are kept moist, so that oxygen can diffuse into the cells.

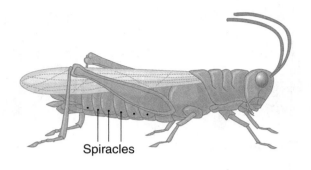

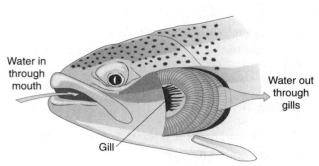

Figure 11.3-3. A grasshopper is an insect that "breathes" through spiracles.

Figure 11.3-4. Gills absorb dissolved oxygen from the water.

Fish

Like humans, fish have a circulatory system that transports oxygen to each cell. Since fish live in water, however, they do not have lungs. Lungs absorb oxygen from the air. Instead, fish have **gills** that absorb oxygen from water. (See Figure 11.3-4.) Gills are made of many folds of tissue filled with blood. Water passes through the mouth of the fish and through the gills, where oxygen is absorbed into the blood. The water then moves out through the gill slits behind the head of the fish.

Frogs

Have you ever seen a baby frog? If you have, it was not on land. Baby frogs are called tadpoles, and they resemble fish. (See Figure 11.3-5.) Tadpoles take in oxygen through gills just like fish do. When a tadpole matures, it grows legs and lungs and comes out onto the land. Vertebrates that begin their lives in water, but move onto land when they mature, are called *amphibians*. Adult frogs can absorb oxygen through their skin as well as through their lungs. Like an earthworm, the skin of an amphibian must be kept moist or it will die.

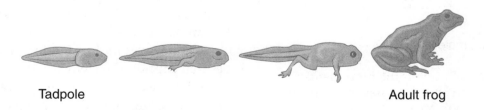

Tadpole

Adult frog

Figure 11.3-5. A tadpole develops into an adult frog.

Activity

Dolphins and whales look like fish, but they are actually mammals. Although they live in water, they do not have gills. They breathe air through their lungs. To learn more about the differences between these sea mammals and fish, visit the Web site *http://www.wdcs.org* and click on "games" on the right side of the page. Select the game "Fish or Mammal." Read the information and then click "Start" at the bottom of the page. Use your mouse to click on the different parts of the animals to learn about their differences. Answer the following questions:

1. Give three examples of cetaceans (sih-TAY-shunz).

2. What are four differences between fish and cetaceans?

Interesting Facts About Respiration

Wouldn't it be nice to have both lungs and gills? Then you could breathe while you were on land as well as in the water. Frogs have both, but not generally at the same time. An adult frog will drown if held under water, and a tadpole will die if it is out of the water. However, there are some animals that actually do have both lungs and gills at the same time. The Australian lungfish has a single lung that enables it to survive when the water dries up. Most of the time, it uses its gills to absorb oxygen from the water, but its lung enables it to survive during dry periods that would kill other fish.

Questions

1. Which process occurs in **all** animals?
   (1) Air is absorbed by the lungs.
   (2) Oxygen enters blood through skin.
   (3) Oxygen diffuses into the cells.
   (4) Oxygen is absorbed through gills.
2. Spiracles are structures that supply air to the cells of
   (1) fish   (3) earthworms
   (2) insects   (4) reptiles

3. Oxygen is absorbed through the moist skin of both
   (1) frogs and snakes
   (2) snakes and birds
   (3) earthworms and frogs
   (4) hydra and humans
4. Which pair of organisms takes in air through their lungs?
   (1) tadpoles and snakes
   (2) snakes and birds
   (3) earthworms and frogs
   (4) hydra and humans

Thinking and Analyzing

1. What two structural differences indicate that tadpoles live in water, while frogs live on land?

2. Identify the structure used to take in oxygen in each of the following animals:
   a. earthworm   c. lion
   b. grasshopper   d. goldfish

11.4

How Do Animals Get Nutrients and Eliminate Wastes?

Objectives

Describe the different methods that animals use to obtain nutrients.

Describe the different methods that animals use to excrete wastes.

Terms

crop: an organ in the digestive tract that stores food

gizzard: an organ in the digestive tract that grinds food

nephridia (neff-RID-ee-uh): excretory organs found in earthworms

proboscis (proh-BAHS-kus): a hollow, tube-like organ used for sucking fluids

Malpighian (mal-PIHG-ee-uhn) **tubules:** excretory organs found in insects

All animals need to take in food to provide their cells with energy and materials for growth and repair. Just like humans, all other animals must convert the food into a form that can be used by the cells. In Chapter 10, you learned about the human digestive system. The human digestive system is a one-way digestive system. Food comes in one end, and the undigested materials go out the other. Many other animals, such as birds, fish, reptiles, and amphibians, have a one-way digestive system as well. Some simpler animals, such as the hydra, have a two-way digestive system. Food comes in and undigested materials go out through the same opening.

Humans excrete wastes three different ways: through the lungs, the skin, and the kidneys. All animals must get rid of the waste materials produced in their bodies. The type of waste products produced

depends in part on the amount of water available to the organism. Each type of animal has a method of excretion that is suited to its diet, metabolism, and environment.

Hydra

Look back at the illustration of a hydra in Figure 11.3-1 on page 338. Each of the tentacles of the hydra contains stinging cells, which can paralyze tiny organisms that brush against it. Once it paralyzes an organism, the hydra uses its tentacles to draw the prey into its mouth. Enzymes digest the food in the stomach cavity of the hydra. Any material that is not digested is pushed back out through the mouth.

Remember that a hydra is only two cells thick. All of its cells are in contact with the watery environment that surrounds it.

Wastes that are produced by the cells of the hydra diffuse out of the cells directly into its watery environment. If you keep hydra in a laboratory, you must change the water often to get rid of these wastes.

Earthworm Digestion

Earthworms are segmented worms. This means that their bodies consist of many repeating rings. (See Figure 11.4-1.) The digestive tract of the worm is a tube that runs through all of these segments.

Earthworms have a one-way digestive system. Food enters the earthworm at one end, and undigested material leaves at the other end. Earthworms live in the soil, which provides them with nutrients. They take in soil through their mouth. Muscular walls push the soil through the esophagus into the **crop,** where it is stored. The soil is then pushed to the **gizzard**, an organ that grinds material into smaller pieces. This ground-up soil then moves through to the intestine, where the nutrients are absorbed into the blood. The materials that cannot be absorbed leave the earthworm's body through the *anus*, an opening at the other end of the earthworm.

Gardeners and farmers love earthworms because they improve the soil. As an earthworm eats its way through the soil, it creates spaces in the soil. These spaces allow air and water to reach the roots of plants. Soil that has many earthworms is usually good for plants. A one-acre plot of good farmland can contain as many as one million earthworms.

Earthworm Excretion

Most of the segments of the earthworm's body have a pair of excretory organs called **nephridia**. (See Figure 11.4-1.) The nephridia are tube-like structures that collect metabolic wastes such as salt and ammonia, and remove them from the earthworm's body. Wastes pass from the nephridia to small openings on the surface of the earthworm's skin. In this way, the nephridia serve the same purpose as the kidneys do in humans and in other vertebrates. However, humans have only two kidneys, while earthworms may have hundreds of nephridia.

Figure 11.4-1. The internal organs of an earthworm.

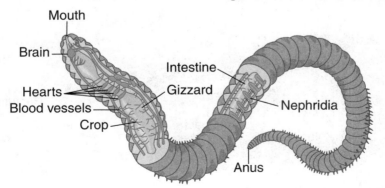

Mouth

Brain

Hearts

Blood vessels

Crop

Intestine

Gizzard

Nephridia

Anus

Figure 11.4-2. The butterfly and the meganosed fly both have a long proboscis for drinking nectar.

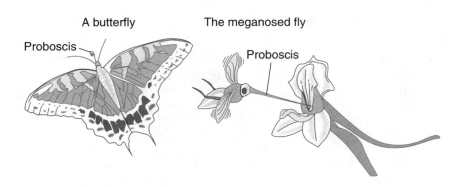

A butterfly

The meganosed fly

Proboscis

Proboscis

Insects

What do insects eat? It depends on the insect. Since there are over one million different species of insects, there are many different insect diets. Some insects eat plants, while others draw in nectar from flowers. Insects may also eat wood, wool, animal wastes, blood, small vertebrates, or even each other. The mouth of the insect is adapted to its diet. Some contain biting parts for chewing plants or for attacking other animals. Other insects, such as butterflies, bees, and the meganosed fly, draw in nectar through a hollow tube called a **proboscis**. (See Figure 11.4-2.) Mosquitoes draw in their food through a hollow tube as well. But because the female mosquito's food is blood, her tube has a sharp end that can pierce the skin of her victim.

Insect Digestion

As you can see in Figure 11.4-3, the digestive system of an insect is similar to the digestive system of an earthworm. Like earthworms,

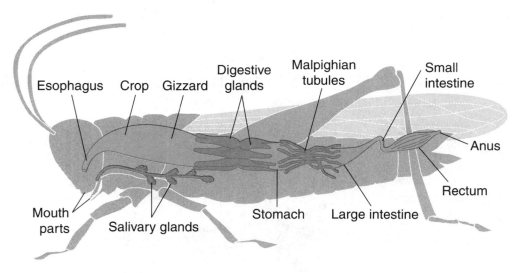

Esophagus   Crop   Gizzard   Digestive glands   Malpighian tubules   Small intestine

Mouth parts   Salivary glands   Stomach   Large intestine   Rectum   Anus

Figure 11.4-3. The digestive system of the grasshopper.

insects have a one-way digestive system. Food passes from the mouth into a crop, where it is stored and partially digested. Once the food leaves the crop, it enters the gizzard. The gizzard has muscular walls, and crushes the food down to a size that can be absorbed into the insect's body. Digestion continues in the stomach. From the stomach, the nutrients enter the blood of the insect, and the blood brings it to the cells. Undigested food passes into an intestine, and then out through the anus.

Insect Excretion

You have already learned about two organs that remove metabolic wastes—kidneys and nephridia. Insects have neither. Both kidneys and nephridia excrete wastes that are dissolved in water. The excretory system of an insect is designed to conserve water. Special structures called **Malpighian tubules** (see Figure 11.4-3) remove wastes from the insect's blood and change them into a material that can be excreted as a solid. These solid metabolic wastes then enter the intestine, where they are eliminated through the anus along with undigested material.

Vertebrates

Fish, amphibians, reptiles, birds, and mammals are vertebrates, the animals most

American Robin   Cardinal

Figure 11.4-4. Compare the beaks of the American robin and the cardinal.

similar to humans. These animals have digestive and excretory systems that are similar to ours. They all have small intestines that deliver nutrients to the blood, and kidneys that remove wastes from the blood.

Just as in insects, however, the mouths of vertebrates are specially adapted to the food sources of each animal. Compare the beaks of the two birds shown in Figure 11.4-4.

Both the cardinal and the robin are common birds found in the eastern United States. The cardinal has a short conical (cone-shaped) beak, which it uses to crack open seeds and nuts. The robin has a longer beak that enables it to dig into the ground to pull out earthworms and insect larvae.

Bird A | Bird B | Bird C

Figure 11.4-5. The heads of a hummingbird, a hawk, and a woodpecker. Which is which?

Figure 11.4-5 shows the heads of three different birds: a hawk that eats mice; a hummingbird that sucks nectar from flowers; and a woodpecker that eats insects that it digs out of the trunk of trees. **Stop and Think:** Which bird is which? **Answer:** The hawk, Bird A, has the sharp, hooked bill used for tearing meat. The woodpecker, Bird B, has the thick, straight bill used for making holes in trees. The hummingbird, Bird C, has the long, thin, curved bill. The beak of a hummingbird serves the same purpose as the proboscis of the butterfly. It allows the hummingbird to gather nectar from flowers.

The shape of the teeth of a mammal indicates the type of food it eats. Lions and tigers are meat eaters. They have large, sharp teeth that are used to tear flesh. Zebras and giraffes have flat teeth that are used to grind down leaves and grasses. **Stop and Think:** What are human teeth designed to do? **Answer:** Humans have both sharp teeth and flat teeth. Humans can eat both plants and animals.

Activity

Go to the Web site
http://www.normanbirdsanctuary.org/beak_adaptations.shtml
Read the descriptions of different bird beaks. Then, in the "Just
for fun" section, click on "Create-a-Bird." Follow the directions
and create your own bird. Then publish your findings by clicking
on the button. If you wish to, you can print the page, and share
your bird with your classmates. Check the "Book of Species" to
see if anyone else created a bird just like yours.

Interesting Facts About Earthworms

How big can an earthworm get? You have probably seen earthworms that
are several inches long. The common earthworms found in the United
States can get as long as 25-30 centimeters, or 10–12 inches. The
largest earthworm ever found, however, was found in South Africa. It was
22 feet (6.7 meters) long!

An American Robin visiting South Africa gets up early to hunt for food.

For more interesting information about earthworms, visit the Web site

http://www.urbanext.uiuc.edu/worms/facts on the Internet.

Questions

1. Which animal has a two-way digestive system?
   (1) earthworm   (3) human
   (2) hydra   (4) grasshopper

2. Which animal has kidneys?
   (1) cat   (3) grasshopper
   (2) earthworm   (4) hydra

3. Complete the analogy: Nephridium is to earthworm, as Malpighian tubule is to
   (1) wasp   (3) hydra
   (2) horse   (4) bluebird

4. Which animal produces solid metabolic wastes?
   (1) hydra   (3) earthworm
   (2) grasshopper   (4) human

5. In this chapter, you read about hydra, earthworms, and grasshoppers. Which of these animals takes in nutrients and excretes wastes?
   (1) only the hydra
   (2) only the earthworm
   (3) only the earthworm and the grasshopper
   (4) the earthworm, the hydra, and the grasshopper

Thinking and Analyzing

1. State two ways that the digestive systems of an earthworm, a grasshopper, and a human are similar.

2. Why is excretion in a grasshopper different from excretion in an earthworm?

3. Describe how excretion takes place in a hydra.

11.5 How Do Animals Respond to Their Environment?

Objectives

Explain how animals respond to changes in temperature.

Explain how animals use chemicals to communicate.

Explain how instincts help animals respond to their environments.

Terms

hypothermia (HY-poh-THUR-mee-uh): a lower-than-normal body temperature

hyperthermia (HY-puhr-THUR-mee-uh): a higher-than-normal body temperature

instincts (IN-stinkts): complex behaviors that are automatic and unlearned

pheromones (FEHR-uh-mownz): chemicals released by an organism that allow it to communicate with other members of its species

In Chapter 10, you learned that humans must maintain a certain body temperature. The enzymes that control human metabolism can work only at temperatures around 99° Fahrenheit (F), which is 37° Celsius (C). Other animals maintain higher body temperatures, and some animals maintain lower body temperatures. If you have a pet cat, you may have noticed that it feels warm when it sits on your lap. Cats have a body temperature of about 102°F, which is 39°C.

Warm-Blooded Animals

Birds and mammals are warm-blooded animals. They must maintain a constant body temperature. This temperature is maintained by the animal's *metabolism—*

the chemical reactions that occur in its body. Most of the time, these reactions produce enough heat to maintain a constant body temperature in cold weather. Sometimes, though, the temperature change in the animal's environment is too extreme, and the animal cannot adjust to it.

Mammals have hair on their bodies. Mammals that live in cold climates have thick fur that protects their bodies from the cold. Since humans do not have thick fur, they can freeze to death in cold conditions. When the outside temperature is too cold, an animal's metabolism may not produce enough heat. When this happens, the animal's body temperature may drop below normal. The condition in which the body temperature drops below the temperature necessary to maintain homeostasis is called hypothermia (the prefix "hypo" means "below," and "therm" refers to "temperature"). If the temperature of an organism is too high, the condition is called **hyperthermia** (the prefix "hyper" means "above"). Either of these conditions can result in death.

Cold-Blooded Animals

Most animals are cold-blooded. Animals that are cold-blooded do not maintain a constant body temperature. Their body temperature changes as the temperature of their environment changes. Cold-blooded animals cannot adapt to changes in temperature as well as warm-blooded animals can. They deal with this problem in several different ways.

Surviving the Winter

Insects are cold-blooded. Some insects move indoors during the winter. Most insects simply die when the weather gets cold. However, their eggs can survive. When the weather gets warm again in the spring, the eggs hatch, and the insects reappear.

Many cold-blooded animals live in areas where the temperature generally does not change very much. Reptiles such as alligators and lizards are common in the southern United States, but cannot survive the winters farther north. Areas such as the Amazon rain forest that have almost the same climate all year round have many different species of cold-blooded animals.

When the internal temperature of a cold-blooded animal drops, its metabolism slows down. All of its body processes tend to become slower as well.

Snakes are found throughout the United States despite the cold winters. During the cold winter months, snakes become inactive. As the air temperature drops and their metabolism slows, they need very little food. They survive on stored food. When their body temperatures are low, snakes move very slowly, which makes them easy prey. To avoid being eaten by predators during winter, snakes find safe places to hide.

Instinct

To survive, animals must find food, water, and shelter, and avoid predators and poisonous wastes. Mammals and birds learn some behaviors from their parents. Other animals need to know what to do, without being taught. Their survival depends completely on their **instincts**, complex behaviors that are inborn, automatic, and unlearned. A stimulus in the environment triggers these behaviors.

The mosquito is an example of an animal that uses instinct to survive. The female uses the blood of humans or other animals to produce her eggs. She is able to find these animals because most of them produce carbon dioxide and heat. The female mosquito can detect the small amounts of carbon dioxide in the air produced when you exhale. She flies toward the source of the carbon dioxide. The female mosquito can also detect the heat produced by a warm-blooded animal, and flies to the source of the heat—your flesh! The female mosquito then inserts a needle-like proboscis into your skin and draws out a small amount of blood. Only female mosquitoes bite. Nobody teaches the female

Figure 11.5-1. A spider and its web.

mosquito to do this and every female mosquito does it. This instinctive behavior ensures the survival of the species.

Many spider species spin complex webs in order to trap their food. Web making by spiders is another example of an instinctive behavior. This behavior is *inborn*—spiders are born with the knowledge of how to build a particular type of web. Figure 11.5-1 shows one type of spider and its web.

Positive and Negative Responses

Many animals are safer when they are in dark places. These animals automatically move away from a source of light. If you shine a light on a common cockroach, it will run to a darker place. Other animals, such as moths, are attracted to sources of light.

Some animals respond to a stimulus by moving toward it, while other animals respond by moving away from it. A movement toward a stimulus is called a *positive* response. A movement away from a stimulus is called a *negative* response.

Animals vary in their responses to other environmental factors. A small bird, called a nuthatch, always climbs down the bark of a tree, while another bird, the brown creeper, always climbs up the tree. *Stop and Think:* To what environmental factor are these birds responding? *Answer:* Gravity! The nuthatch moves toward Earth's gravity, while the brown creeper moves away from it.

Chemical Messages

Have you ever wondered how Mr. and Mrs. Insect find each other and produce their young? Insects produce chemicals called **pheromones**, which enable them to find each other. Pheromones are chemicals released by an organism into its environment that allow it to communicate with other members of its species. The male moth, for example, senses the pheromones of the female in the air and flies toward it. When it finds the female producing the pheromones, it mates with her. Scientists hope to be able to manufacture these pheromones and use them to attract and trap the males. This will prevent harmful insects from mating, and thus control their populations.

Interesting Facts About Ant Pheromones

You have just learned that insects use chemicals called pheromones to communicate with each other. Finding a mate is not the only way that insects use these chemicals. Have you ever seen a long line of ants? Ants release pheromones that guide other ants to a source of food. As long as the food supply lasts, the ants continue to release these pheromones. When the food runs out, the ants no longer release the chemicals, and the trail disappears.

To learn more about pheromones, visit the Web site *http://biology-pages.info*

Click on the letter "P" and scroll down to the word "pheromones." Then click on the word "insect."

Questions

1. A small flatworm, called a planarian, moves away from light. This movement is best described as a
   (1) positive stimulus
   (2) negative stimulus
   (3) positive response
   (4) negative response

2. Which animal maintains a constant body temperature?
   (1) grasshopper   (3) gorilla
   (2) earthworm   (4) goldfish

3. Which behavior could be considered an instinct?
   (1) A child eats his breakfast.
   (2) A crow learns how to open a garbage can to get food.
   (3) A spider builds a web to catch insects.
   (4) A blue jay returns to the same bird feeder every morning.

Thinking and Analyzing

1. When the *Titanic* sank in 1912 after colliding with an iceberg, more than 1,000 people died in the 31°F (−1°C) water, even though they were wearing life jackets. What killed these people, if they did not drown?

2. How do insects know where to find a mate?

3. The American robin is a bird that can be seen throughout the northern United States from March until November. Robins eat earthworms and insects, which they find in the soil. Why do these birds fly south during the winter months, and return in the spring?

4. In cold weather, a bird's metabolism increases, while a snake's metabolism decreases. Explain why these two animals respond differently to a change in temperature.

Review Questions

Term Identification

Each question below shows two terms from Chapter 11. One of the terms is defined.
(1) Choose the term that matches the definition.
(2) Describe how the two terms are different. Following each term is the section (in parenthesis) where the description or definition of that term is found.

1. *Vertebrate (11.1) — Invertebrate (11.1)*
   An animal that does not have a backbone

2. *Exoskeleton (11.2) — Endoskeleton (11.2)*
   The hard outer covering found on arthropods

3. *Diffusion (11.3) — Gills (11.3)*
   Structures in fish that remove dissolved oxygen from water

4. *Crop (11.4) — Gizzard (11.4)*
   An organ in the digestive tract that stores food

5. *Nephridia (11.4) — Malpighian tubules (11.4)*
   Excretory organs found in insects

6. *Spiracles (11.2) — Proboscis (11.4)*
   A hollow tube-like organ used for gathering fluids

7. *Hypothermia (11.5) — Hyperthermia (11.5)*
   A lower-than-normal body temperature

8. *Instincts (11.5) — Pheromones (11.5)*
   Complex behaviors that are automatic and unlearned

Multiple Choice (Part 1)

Choose the response that best completes the sentence or answers the question.

1. Which of the following animals is an invertebrate?
   (1) bird   (3) insect
   (2) fish   (4) reptile

2. Which group contains the **smallest** number of different species?
   (1) mammals   (3) invertebrates
   (2) vertebrates   (4) reptiles

3. Which animal has an endoskeleton?
   (1) hydra   (3) grasshopper
   (2) earthworm   (4) snake

4. Complete the analogy: Endoskeleton is to bone as exoskeleton is to
   (1) chitin   (3) arthropod
   (2) insect   (4) skin

5. Worms, hydra, and jellyfish are examples of animals that have
   (1) exoskeletons   (3) vertebrae
   (2) endoskeletons (4) soft bodies

6. Which process occurs in **all** animals?
   (1) Air is absorbed by the lungs.
   (2) Wastes are returned to the environment.
   (3) Water is released through the skin.
   (4) Excretion is controlled by the kidneys.

7. Complete the analogy: Spiracles are to lungs as Malpighian tubules are to
   (1) heart   (3) intestines
   (2) kidneys   (4) gizzards

8. Nephridia are involved in the life function called
(1) reproduction (3) nutrition
(2) respiration (4) excretion

9. Functions carried out by the gills of fish, the lungs of humans, and the skin of earthworms illustrate the idea that
(1) all animals carry out the same life functions in the same way
(2) all animals contain the same organs to carry out life functions
(3) different animals may have different organs that carry out the same life functions
(4) not all animals need to carry out life functions

10. Which organ is involved in a different life process from the other three?
(1) gizzard (3) intestine
(2) stomach (4) gill

11. The hydra does not have an anus because it
(1) has a one-way digestive system
(2) has a two-way digestive system
(3) is an invertebrate
(4) lives in water

12. Which animal has a lower body temperature in the winter than it does in the summer?
(1) human (3) bullfrog
(2) horse (4) bluebird

13. Nest building in birds is called an instinct because
(1) it is necessary to produce the next generation
(2) birds learn how to build nests by watching other birds do it
(3) nest building is automatic and unlearned behavior
(4) all birds do it

14. A pheromone is best described as a
(1) chemical stimulus
(2) visual stimulus
(3) learned response
(4) reproductive organ

Thinking and Analyzing (Part 2)

1. Why is an earthworm considered an invertebrate, but **not** an arthropod?

2. Hydra obtain their oxygen by a process called diffusion. What is diffusion?

3. Oxygen enters the human body through the lungs. Name two organs in two other animals that perform the same function as the lung. Be sure to name both the animal and the organ.

4. Match each of the organs in the first column with the appropriate animal and life function in the second and third columns. Give just one correct combination for each organ. Animals and life functions may be used more than once (for example, one possible correct answer is Organ: gizzard; Animal: earthworm; Life function: nutrition).

Organ	Animal	Life Function
Tentacle	Hydra	Excretion
Spiracle	Earthworm	Respiration
Nephridia	Grasshopper	Nutrition
Gill	Fish	
Gizzard	Cat	
Malpighian tubules		
Kidney		

Chapter Puzzle (*Hint:* The words in this puzzle are terms used in the chapter.)

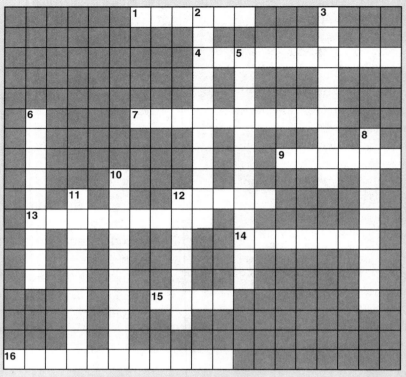

Across

1 hyperthermia is defined as a ___-than-normal body temperature

4 chemicals released by an organism that allows it to communicate with other members of its species

7 insects, lobsters, and spiders, but not worms

9 the hard material that makes up the exoskeleton

12 structures in fish that remove dissolved oxygen from water

13 excretory organs found in earthworms

14 insects have excretory organs called Malpighian ___

15 an organ in the digestive tract that stores food

16 the outer skeleton found on arthropods

Down

2 a lower-than-normal body temperature

3 a hollow, tube-like organ used for drawing in fluids

5 the system of bones found inside an animal

6 behaviors that are automatic and unlearned

8 oxygen moves through a cell membrane by a process called ___

10 an animal with a backbone

11 tiny openings in the bodies of insects that take in air

12 an organ in the digestive tract that grinds food

Chapter 12

Plants

Contents

Sunflowers turn to face the sun.

What Is This Chapter About?

In Chapter 11, you learned how different kinds of animals perform life functions and maintain homeostasis. In this chapter, you will investigate the organisms in a completely different kingdom—the plant kingdom. You will find that although plants and animals share the same life functions, they carry them out using very different kinds of organs.

In this chapter you will learn:

1. All plants must carry out life process to maintain homeostasis.

2. The main organs of plants are the stems, roots, leaves, and flowers.

3. Leaves are adapted to carry out photosynthesis.

4. Stems support plants and transport water and nutrients.

5. Roots anchor plants and absorb water.

6. Flowers contain the reproductive organs of plants.

7. Plants respond to changes in their environment.

Science in Everyday Life

Have you ever grown a plant? What must you do to keep a plant alive? Plants need water, air, and light. Some plants need a large amount of water, while others require very little. Some plants grow best in direct sunlight, while others do better in shade. No matter what kind of plant you grow, understanding its life functions will make you a better gardener.

Internet Sites:

http://www.biology4kids.com/files/plants_main. html Everything you need to know about plant biology is found at this informative Web site.

http://www.mbgnet.net/bioplants/ Go through each of the links on the left side of the page to learn important information about plants. Be sure to look at all the great pictures and watch the videos.

12.1   What Are the Parts of a Plant?

Objectives

Identify the four main parts of a plant.

Identify the functions of the stems, roots, leaves, and flowers.

Terms

root: the plant organ that anchors the plant and absorbs water

leaf: the plant organ where most photosynthesis takes place

chlorophyll: a green chemical plants need for photosynthesis

stem: the plant organ that supports the plant and transports materials

flower: the part of the plant that contains its reproductive organs

---

Plant Parts

Plants are living things. Like all living things, plants carry out life functions, such as respiration, nutrition, reproduction, and response to stimuli. Because plants are multicellular, they must also transport materials to and from their cells. Plants contain organs that enable them to carry out these life functions.

Figure 12.1-1 shows a vinca plant. You can see three plant organs—the leaves, the stem, and the flower. *Stop and Think:* What organ is *not* shown in the picture? *Answer:* The roots are not shown. Roots are parts of a plant that are usually found below ground level. Roots, stems, and leaves will be discussed in greater detail in Lessons 12.2 to 12.4. For now, let's take a brief look at what each of these organs does.

Figure 12.1-1. A vinca plant has roots, stems, leaves, and flowers.

Roots

Roots grow underground and connect the plant to the soil. (See Figure 12.1-2.) Plants do not move from place to place because their roots anchor them. During a bad storm, however, a strong wind can pull a plant out of the soil. When that happens, we say that the plant has been "uprooted." Although anchoring a plant is an important function of the roots, they do a lot more than just keep the plant in the ground!

The soil contains water and other nutrients that the plant needs to stay alive. The roots spread out through the soil to find and absorb these materials. Some plants, such as carrots, radishes, and sweet potatoes, have roots that can store both food and water.

Leaves

What color are leaves? Most leaves are green because they contain a green compound called **chlorophyll**. Chlorophyll is needed for photosynthesis, the process that makes food for a plant. (See Lesson 8.3.) Most photosynthesis takes place in leaves.

In photosynthesis, a plant uses sunlight, water, and carbon dioxide to produce sugar and oxygen. The water comes from the roots; the leaves provide the rest. Leaves absorb sunlight and exchange carbon dioxide and oxygen with the environment.

Stems

The roots of a plant absorb water that is needed by the leaves for photosynthesis.

Stop and Think: How does water get from the roots to the leaves?

Answer: The **stem** of the plant is the organ that carries water and other dissolved materials to and from the roots and the upper parts of a plant. The stem has another function: it supports the plant. We could say that the functions of the stem in plants are similar to those of both the skeletal and circulatory systems in animals.

Most stems, like the stem shown in Figure 12.1-2, are aboveground. Some plants, however, have underground stems as well. Tulips, lilies, onions, and potatoes, for example, contain underground stems that store food to help them survive through the winter.

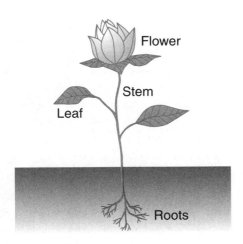

Figure 12.1-2. The parts of a plant.

Flowers

Flowers are the reproductive parts of a plant. They usually contain both the male and the female sex organs. Figure 12.1-3 shows the parts of a flower. Reproduction in a flower produces a seed, which can produce a new plant. The ovary develops into the fruit of a plant. A fruit is a seed container. You will learn more about reproduction in both plants and animals next year.

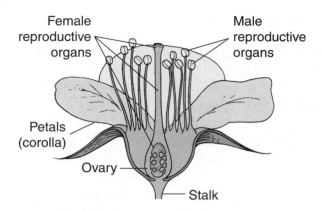

Figure 12.1-3. Parts of a flower.

Activity

Visit the Web site *http://www.mbgnet.net/bioplants/parts.html* on the Internet.
Read the page, scroll down, and then click on the vegetables to play the Play a Plant Parts Game. Which foods surprised you?

Interesting Facts About Plant Parts

Do you eat your vegetables? Exactly what are vegetables? Scientists define vegetables as foods that come from the growing parts of a plant. These usually include the leaves, stem, and roots. Asparagus is the stem of a plant. Potatoes, garlic, and onions are underground stems. Carrots and sweet potatoes are roots. Have you ever eaten cauliflower? As the name suggests, cauliflower is a flower. Broccoli and artichokes are also vegetables that come from the flower of the plant. Vegetables do *not* include fruits and seeds. Strictly speaking, many of the foods you consider vegetables are actually fruits! If food contains seeds, biologically speaking, it is a fruit. Cucumbers and peppers are both fruits.

Questions

1. Water enters the plant through the
   (1) flowers   (3) roots
   (2) fruits   (4) stem

2. The two main functions of the stem are
   (1) transport and reproduction
   (2) support and photosynthesis
   (3) reproduction and taking in water
   (4) support and transport

3. Most photosynthesis takes place in a plant's
   (1) leaves   (3) stems
   (2) roots   (4) flowers

4. Both male and female reproductive organs can be found in a plant's
   (1) leaves   (3) stems
   (2) roots   (4) flowers

Thinking and Analyzing

1. Which of the four main parts of a plant can be removed without killing the plant? Explain your answer.

2. A head of lettuce is not really a head and an ear of corn is not really an ear. What part of a lettuce plant is the head? What part of a corn plant is the ear?

3. Explain the following statement: Plants need to drink, but they don't need to eat.

4. The diagram at the right shows a bean plant.
   Copy the table below into your notebook. Complete it by identifying the *three* structures labeled *A, B,* and *C.* Identify one function of each structure.

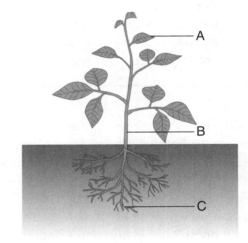

Letter	Plant Structure	Function of Structure
A		
B		
C		

How Do Plants Obtain Energy?

Objectives

Identify photosynthesis as the source of food in plants.

Identify some parts of a leaf.

Explain how these parts work together to perform photosynthesis.

Terms

chloroplasts: structures in plant cells where photosynthesis occurs

stomata: openings in leaves that permit the exchange of gases

palisade layer: the region within a leaf where most photosynthesis occurs

---

All living things require energy to stay alive. Animals get all of their energy from plants. Even animals that do not eat plants get their energy from plants. A gazelle is a plant-eating animal. It uses some of the energy it gets from plants to carry on life functions. The energy it doesn't use is stored in its body. When a cheetah eats the gazelle, it obtains that stored energy. The cheetah obtains its energy from the plant, through the gazelle. *Stop and Think:* Where does a plant get its energy? *Answer:* A plant gets its energy from the sun. The process of photosynthesis uses the sun's energy to produce food for the plant, and for any animal that eats the plant.

Photosynthesis

Recall that photosynthesis is a chemical reaction, with the equation:

Carbon dioxide + water + energy → glucose + oxygen

If *you* mix carbon dioxide with water, all you can form is seltzer! In Lesson 9.2, you learned that plant cells have special organelles called **chloroplasts**. Chloroplasts contain *chlorophyll*, the green compound that enables plants to produce glucose (a sugar) from carbon dioxide and water. As you can see from the equation, photosynthesis requires energy. The energy for photosynthesis comes from sunlight. (The prefix *photo-* means "light," and *synthesis* means "putting together.") Plants contain special organs that bring together all of the necessary materials for photosynthesis: carbon dioxide, water, chlorophyll, and sunlight. *Stop and Think:* What are these organs called?

Answer: Leaves.

Leaves

Look at the leaf in Figure 12.2-1. Most leaves are wide and flat, so that they can gather the maximum amount of sunlight. Although leaves are thin, they contain several layers of cells. Have you ever noticed that the top of a leaf looks different from the bottom? The top has a waxy coating that prevents the loss of water. The bottom contains many tiny openings called **stomata**, or *stomates*. A single opening is called a *stoma*. Carbon dioxide, oxygen, and water vapor move into and out of the leaf through the stomata. Leaves contain veins that bring water to their cells and carry glucose to other parts of the plant. (See Lesson 12.3.)

Water, carbon dioxide, chlorophyll, and sunlight all come together in the **palisade layer**. The cells in this layer contain large numbers of chloroplasts, and this is where most of the photosynthesis takes place. The sugars produced by photosynthesis store energy that came originally from the sun.

Using Stored Energy

Plant cells, like all other cells, use energy to stay alive. *Cellular respiration* occurs in both animal cells and plant cells. During cellular respiration in plants and animals, glucose reacts with oxygen to produce water and carbon dioxide. This reaction releases the energy that was stored in the glucose. (See Lesson 8.3.) Cellular respiration can be summarized by the equation:

Glucose + oxygen → water + carbon dioxide + energy

Both photosynthesis and respiration occur in all plants. Photosynthesis occurs only in the daytime, while respiration occurs all the time.

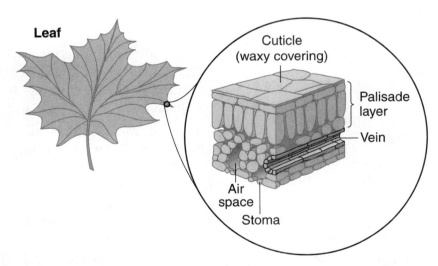

Figure 12.2-1. Leaves are very thin; the image in the circle is greatly magnified. The cuticle is the top surface of the leaf, while the stomata are on the underside.

Activity

Visit the Web site *www.sciencemadesimple.com/leaves.html*
Read the material about autumn leaves, and try the word
scramble at the end of the passage.

Interesting Facts About Leaves

People come from all over the world to spend autumn in New England. One of the reasons they come is to observe the changing colors of the leaves. New England is home to a wide variety of *deciduous trees* — trees that lose their leaves in the fall. Before the leaves drop from the trees, they turn brilliant shades of orange, yellow, and red. Where did these colors come from? Actually, they were in the leaves all summer long! During the spring and summer, the leaves contain so much green chlorophyll that the other colors can't be seen. As the tree prepares for winter, the chlorophyll disappears, revealing the other beautiful colors in the leaves.

Questions

1. Photosynthesis requires all of the
   following **except**
   (1) chlorophyll   (3) carbon dioxide
   (2) glucose   (4) water

2. The main function of leaves is to
   (1) support the other parts of the plant
   (2) manufacture food for the plant
   (3) absorb water from the soil
   (4) produce new plants

3. Gases enter and exit a leaf through the
   (1) stomata   (3) chloroplasts
   (2) veins   (4) palisade layer

4. Photosynthesis requires a constant
   supply of energy. This energy comes from
   (1) chlorophyll   (3) sunlight
   (2) glucose   (4) carbon dioxide

5. Which statement is true?
   (1) Photosynthesis occurs only during
   the day and respiration occurs only at
   night.
   (2) Photosynthesis occurs all the time
   and respiration occurs only during the
   day.
   (3) Photosynthesis occurs only at night
   and respiration occurs all the time.
   (4) Photosynthesis occurs only during
   the day and respiration occurs all the
   time.

Thinking and Analyzing

Base your answers to questions 1–3 on the diagram to the right and on your knowledge of science. The diagram shows the sun and a green plant.

1. Identify *one* part of the plant that carries on photosynthesis.

2. In addition to sunlight and chlorophyll, what are the *two* materials that a plant needs to carry on photosynthesis?

3. Identify *one* product that results from photosynthesis.

Base your answers to questions 4 and 5 on the paragraph below and your knowledge of science.

Global warming is an issue that is constantly in the news. Most scientists believe that increased amounts of carbon dioxide in the air are causing Earth's temperature to increase. Carbon dioxide is produced when we burn fuels such as coal, oil, and gasoline.

Some scientists believe that destroying trees also contributes to global warming. Trees and other green plants are nature's way of removing carbon dioxide from the air.

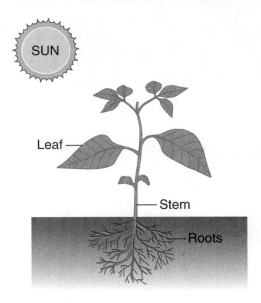

4. What plant process removes carbon dioxide from the air?

5. What action might people take to limit the amount of carbon dioxide they put into the air?

12.3 How Do Plants Transport Nutrients?

Objectives

Describe how materials are transported inside a plant.

Distinguish the movement of materials in the xylem from the movement of materials in the phloem.

Describe the role of the stem and the roots in transporting nutrients through the plant.

Terms

herbaceous (huhr-BAY-shuhs) **stem:** green stems found on non-woody plants

xylem (ZY-luhm): the tissue that carries materials upward through a plant

transpiration (tranz-puh-RAY-shun): the escape of water molecules from the leaves of the plant

phloem (FLOH-uhm): the tissue that carries materials downward through a plant

vascular (VASS-cue-luhr) **bundles:** the tissue in herbaceous stems that contains the xylem and phloem

root hairs: hair-like structures that absorb water into the root

---

The redwood trees in California can grow to be 300 feet tall. For photosynthesis to take place, water from the soil must reach the leaves at the top of the tree. To keep the root cells alive, sugar that is made in the leaves must reach the roots. Plants have a transport system that is capable of moving materials over great distances. The stem of the plant contains specialized tubes that carry materials upward, and other specialized tubes that carry materials downward.

Stems

Figure 12.3-1 shows a maple tree. Where is the stem of the plant? Some plants have *woody* stems. In these plants, a hard, rough protective covering surrounds the stem. The trunk of a tree is actually a woody stem. Other plants, such as tulips, daffodils, dandelions, and almost all houseplants have soft green stems. These are called **herbaceous stems**. Whether the stem is herbaceous or woody, it contains tubes that transport materials. The arrows in Figure 12.3-1 represent the movement of materials through the stem.

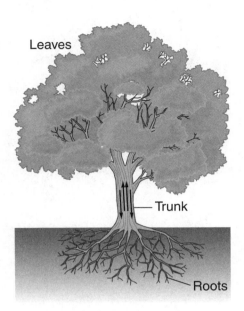

Figure 12.3-1. A maple tree has a woody stem.

Xylem

The tissue that carries materials *up* the plant is called the **xylem**. The xylem forms tubes that begin in the roots and continue through the stem, eventually reaching the leaves. *Stop and Think:* What is the main material that moves through the xylem? *Answer:* Water moves through the xylem, from the roots to the rest of the plant. When water reaches the leaves, some of it may escape through the stomata, into the air. **Transpiration** is the escape of water molecules from the leaves of the plant. As the water escapes from the leaves, fresh water is pulled up through the xylem. The roots, xylem, and stomates work together to maintain a constant supply of water throughout the plant.

Figure 12.3-2 compares the cross sections of two different types of stems. Notice that in the woody stem the xylem appears in rings, moving out from the center of the stem. New xylem forms constantly, causing the stem to become wider. The number of rings of xylem inside the stem indicates the age of the tree. The tree in Figure 12.3-2 is three years old. The herbaceous stem of a corn plant has xylem as well, but it is arranged in bundles, not in rings.

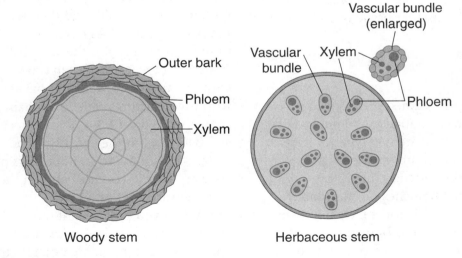

Figure 12.3-2. Cross sections of a woody stem and a herbaceous stem.

Phloem

Phloem is the name of the tissue that carries materials *down* the plant. In woody plants, the phloem is located just inside the outer layer of bark. In herbaceous plants, phloem is located in the same bundles of tissue that contain the xylem. (See Figure 12.3-2.) **Stop and Think:** What materials move through the phloem? **Answer:** Dissolved sugars and other nutrients produced by the leaves move through the phloem to the rest of the plant. Trees that lose their leaves in the fall cannot manufacture food during the winter. They store nutrients in a fluid called sap. The tree uses the sap for energy until the leaves reappear in the spring. Recall that some plants store nutrients in *underground* stems. In these plants, as in trees, the nutrients are carried by the phloem, from the leaves, through the stems.

The bundles of tissue shown in Figure 12.3-2 are called **vascular bundles**. The word "vascular" is related to the word "vessel." Vascular tissue carries materials through both plants and animals. In plants, vascular tissue includes the xylem and phloem. In many animals, vascular tissue includes veins, arteries, and capillaries.

Roots

You have already learned that roots have two main functions—to anchor a plant and to absorb water. There are two types of roots: fibrous roots and taproots. (See Figure 12.3-3.) Taproot systems contain one large, central root that branches out into many smaller roots. Fibrous roots have no larger central root. Their roots branch out in all directions from the bottom of the stem. All roots end in a hard covering called a *root*

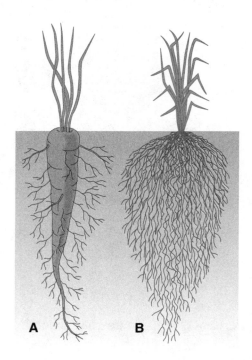

Figure 12.3-3. (A) A taproot and (B) a fibrous root.

cap. The root cap protects the root as it pushes its way through the soil. (See Figure 12.3-4.)

If you pull a small plant out of the ground, roots and all, and then replant it, the plant usually does not survive. The top of the plant wilts, and the stem collapses from lack of water. Why can't the roots continue to absorb water? Roots are covered with tiny structures called **root hairs**, as shown in Figure 12.3-4. Root hairs are thin and delicate, but they greatly increase the surface area of the root. Root hairs provide a large amount of contact between the root and the water and other nutrients in the soil. When you pull a plant out of the ground, you often destroy the root hairs. The roots that remain are not able to absorb enough water to keep the plant alive.

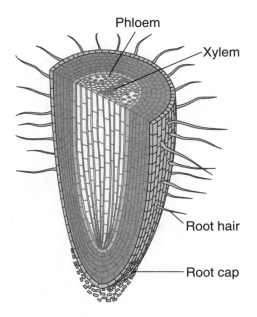

Phloem

Xylem

Root hair

Root cap

Figure 12.3-4. Root hairs increase the surface area of the root so it can absorb water.

Stop and Think: Water enters the root through the root hairs. How does the water move through the root to the stem? *Answer:* Notice in Figure 12.3-4 that the root contains xylem and phloem, just as the stem does. Water moves through the cells into the xylem, where it begins its journey upward. Why is there also phloem in the root? Like all living cells, the cells in the root need nutrients to stay alive. The phloem transports these nutrients down to the root cells. Some taproots, such as in carrots and rhubarbs, can store these nutrients, which feed the plant during the winter.

Interesting Facts About Sap

Have you ever poured maple syrup on your pancakes or French toast? Maple syrup is made from the sap of the sugar maple tree. A simple wooden tube is driven through the bark of the tree, into the phloem. A bucket placed under the tube collects the sap. Sap is taken only from mature, healthy sugar maples, which are not harmed by the process. When the sap comes out of the tree it is about 2 percent sugar, and watery—not syrupy. The sap is boiled until it becomes the thick, sweet product known as maple syrup. It takes more than 40 quarts of sap to produce just one quart of maple syrup.

Activity

Get a stalk of celery that has some leaves at the top. Remove the bottom of the stalk by cutting it straight across. Place the cut end of the celery into half a glass of water containing about 4 drops of red food coloring and 4 drops of blue food coloring mixed together. Allow the celery to sit in the colored water for about 2 hours.

Remove the celery from the colored water and cut off the bottom *above the water level.* Examine the end of the celery stalk. What do you observe? Draw the results in your notebook. Cut the celery again about one inch higher and examine it again. Record the results. Repeat the procedure to determine how high the colored water has risen in the celery stalk.

Change a variable and do the experiment again. What happens if you leave the celery in the colored water overnight? How might the results differ if there were no leaves at the top of the celery? How would blowing on the leaves with a hair dryer affect the results? Would wilted celery behave differently from fresh celery? Apply one of these changes to the original procedure or make up your own. Determine how your change affects the way that the colored water travels in the celery stalk. Share your results with your class.

Questions

1. Xylem and phloem are found in
   (1) only the stems of plants
   (2) only the roots of plants
   (3) both the stems and roots of plants
   (4) neither the stems nor roots of plants

2. Water and dissolved minerals are absorbed by a plant and travel
   (1) up the phloem
   (2) up the xylem
   (3) down the phloem
   (4) down the xylem

3. Water enters the root through the
   (1) root hair
   (2) root cap
   (3) xylem
   (4) phloem

4. Phloem enables the plants to
   (1) digest food
   (2) carry out cellular respiration
   (3) excrete wastes
   (4) transport food

Thinking and Analyzing

1. How do phloem and xylem differ?

2. There are two types of stems and two types of roots. What are they?

3. Water that enters the roots eventually returns to the roots, bringing nutrients that keep the cells in the roots alive. Describe the path that water takes, from when it first enters the roots to when it returns to the roots.

12.4 How Do Plants Respond to Changes in Their Environment?

Objectives

Give examples of three types of tropism in plants.

Explain how growth can cause a plant to bend toward or away from a stimulus.

Describe how plants respond to the changing seasons.

Terms

tropism (TROH-piz-uhm): the growth of a plant toward or away from a stimulus

phototropism (FOH-toh-TROH-piz-uhm): the growth of a plant toward or away from light

geotropism (JEE-oh-TROH-piz-uhm): the growth of a plant toward or away from the pull of gravity

thigmotropism (THIG-moh-TROH-piz-uhm): the response of a plant to touch

hydrotropism (HIGH-droh-TROH-piz-uhm): the growth of a plant toward or away from water

annuals: plants that live less than one year

dormant: inactive

perennials (puhr-EN-ee-uhlz): plants that live for several years

---

Plants do not move from place to place. They do not hunt for food, water, or shelter. When you watch a plant from day to day, you may see some growth but little other change. Yet plants do respond to their environment. Flowers open and close, stems bend, roots dig, leaves fall.

Tropisms

You may have observed that houseplants bend toward a light source. This type of movement is called a **tropism**. A tropism is the growth of a plant toward or away from a stimulus. Tropisms are named for the type of stimulus that causes them. The bending of a plant toward light is called **phototropism** because *photo* means "light." (See Figure 12.4-1.)

Tropisms involve growth. How can growth cause a plant stem to bend? A plant stem bends when one side of the stem grows more, that is, faster, than the other side. *Stop and Think:* Which side of the stem grows faster, the side facing the light source, or the side facing away from it? *Answer:* The side *away* from the light source grows faster. Tropisms that move the plant toward the

Figure 12.4-1. A plant bends toward a source of light.

stimulus are called *positive* tropisms. When a plant bends toward the light, it is a positive tropism. There are also negative tropisms. How is a plant affected by gravity? One part of the plant moves downward, while another part moves upward. The movement toward or away from the pull of gravity is called a **geotropism**. Roots show a positive geotropism, while stems show a negative geotropism.

Have you ever seen the way a vine wraps around a fence? Many plants, including grapes, morning glories, and sweet peas, use fences, string, or other plants for support. When the plant touches a suitable object, it reacts by coiling around it. The plant coils because the side *away* from the object grows faster than the side *touching* the object. The response of a plant to touch is called **thigmotropism**.

The roots of a plant need to absorb water. In many plants, the roots can detect the presence of water and grow toward the water. The movement of the roots toward water is called **hydrotropism**.

Adapting to Seasonal Change

A garden in May can have hundreds of flowers, but in January many gardens are just bare ground. Even if the gardener puts in no new plants, many of the previous year's plants reappear the next spring. How does this happen?

Annuals

Many flowering plants are **annuals**. These are plants that live less than one year; they are killed by the cold winter weather. However, annuals produce seeds that remain **dormant** during the winter. The word *dormant* comes from a Latin word that means "sleeping." A dormant seed or plant is inactive. It is not growing and is using very little energy. The seeds do not begin to produce new plants until the soil becomes warm, usually in the spring. Marigolds and petunias, shown in Figure 12.4-2 on page 376, are annuals.

When annuals reappear where they were planted the previous year, the new plants are the offspring of the previous year's plants. They are not the same plants, and may even show different colors and shapes from those of their parents.

Perennials

Other plants in the garden are **perennials**. They come back every year. During the winter, the perennial does not die. The part of the plant that is aboveground may disappear, but the part below the ground is

Marigolds   Petunias

Figure 12.4-2. Two common annuals, marigolds and petunias.

still alive. Perennials store nutrients in their roots or stems. They remain dormant during the winter, but when the soil becomes warm, they use those nutrients to produce new growth. Tulips, daffodils, lilies, and chrysanthemums are perennials. (See Figure 12.4-3.)

Deciduous Trees

Deciduous trees, and many small bushes, such as rose plants, lose their leaves in the fall and become dormant. They use very little energy during the winter. Like herbaceous perennials, trees and bushes store enough food to help them grow again in the spring. The stored food provides the energy that enables the plants to produce new growth and develop new leaves. Once the leaves have developed, the plant can produce its own food through photosynthesis.

Tulips   Daffodils   Lillies   Chrysanthemums

Figure 12.4-3. Perennial garden plants.

SKILL EXERCISE—*Interpreting a Graph*

A gardener performs an experiment growing three different types of plants in equal amounts of soil. Each plant is 10 centimeters tall at the beginning of the experiment. The three plants are given 4 milliliters of water every day for 20 days. The results of the experiment are shown on the graph.

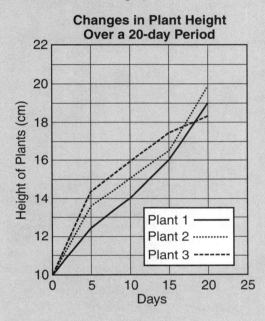

Changes in Plant Height Over a 20-day Period

1. Which plant was the tallest at the end of the 20-day period?

2. Other than at the beginning of the experiment, on what day were plant 2 and plant 3 the same height?

3. Which plant grew at the slowest rate from day 0 to day 5?

4. During which time interval did plant 1 become taller than plant 3?

   (1) days 0–5

   (2) days 5–10

   (3) days 10–15

   (4) days 15–20

Interesting Facts About Plants: Responding to Touch

It seems only fair, that with so many animals eating plants, there would be at least a few plants that eat animals! The pitcher plant, sundew, and Venus flytrap are examples of plants that eat animals. The pitcher plant drowns insects, and the sundew traps them on hairs covered with a sticky fluid. The most interesting, though, may be the Venus flytrap, illustrated below. The plant has harmless white flowers, but the tips of the leaves develop into deadly traps. The figure shows how one of these traps has captured a fly. Each trap contains a few small hairs on each side. When an insect brushes against two of them, the trap snaps shut. The plant digests the insect, and the trap reopens a few days later. You can purchase your own Venus flytraps at many garden stores, or through the Web.

A Venus flytrap.

Questions

1. The stem of the plant to the right is bending toward the window. This demonstrates a
   (1) negative phototropism
   (2) positive phototropism
   (3) positive geotropism
   (4) negative geotropism

2. The stem bends because the cells in the part of the stem that faces **away** from the window
   (1) grow slower than the cells that face toward the window
   (2) grow faster than the cells that face toward the window
   (3) begin to die due to the lack of light
   (4) become smaller due to the lack of light

3. Which parts of a plant are most likely to show a **positive** geotropism?
   (1) the roots   (3) the stems
   (2) the leaves   (4) the flowers

4. Cherry trees are deciduous trees that grow well in many parts of the country.

During the winter, we would expect cherry trees to
(1) continue to produce food through photosynthesis
(2) use energy from the sun to produce fruit
(3) become dormant
(4) die

Thinking and Analyzing

Base your answers to questions 1–3 on the following reading passage and on your knowledge of science.

Growing Grapes

Many people in New York and California grow their own grapes. Grape plants require sunlight, a lot of space, and a fence. A grape plant is a vine, a type of plant that uses other plants for support. A special type of fence called a trellis is usually used to support the grape plant. The stems of the grape plant have rope-like structures called tendrils. When the tendrils come in contact with something solid, such as a fence or another plant, they wrap tightly around it. In the winter, when the plants become dormant, the grape leaves fall from the plant, but the tendrils remain connected to the fence.

1. When the tendrils touch the fence, they wrap around it. What name is given to this type of tropism?

2. When the tendrils wrap around the fence, which part grows faster, the side in contact with the fence, or the side **not** in contact with the fence? Explain your answer.

One type of trellis.

3. What other tropism would you expect to observe in grape plants? Explain your answer.

Review Questions

Term Identification

Each question below shows two terms from Chapter 12. One of the terms is defined.
(1) Choose the term that matches the definition.
(2) Describe how the two terms are different.
Following each term is the section (in parentheses) where the description or definition of that term is found.

1. *Root (12.1) — Stem (12.1)*
   A plant organ that anchors the plant and absorbs water

2. *Leaf (12.1) — Flower (12.1)*
   The part of a plant that contains its reproductive organs

3. *Chlorophyll (12.1) — Chloroplasts (12.2)*
   Structures in plant cells that are the sites of photosynthesis

4. *Stomata (12.2) — Palisade layer (12.2)*
   The region in a leaf where most photosynthesis occurs

5. *Xylem (12.3) — Phloem (12.3)*
   The tissue that carries materials upward through a plant

6. *Root hairs (12.3) — Vascular bundles (12.3)*
   The tissue in herbaceous stems that contains the xylem and phloem

7. *Phototropism (12.4) — Geotropism (12.4)*
   The growth of a plant toward or away from the pull of gravity

8. *Hydrotropism (12.4) —Thigmotropism (12.4)*
   The response of a plant to touch

9. *Annuals (12.4) — Perennials (12.4)*
   Plants that live for many years

Multiple Choice (Part 1)

Choose the response that best completes the sentence or answers the question.

1. Which observation of a plant supports the inference that photosynthesis can take place?
   (1) a strong, sweet smell
   (2) a dry, rough texture
   (3) a green color
   (4) a smooth stem

2. Complete the analogy: Phloem is to down, as ___ is to up.
   (1) leaf   (3) chlorophyll
   (2) xylem   (4) stomata

3. Most carbon dioxide enters a plant through its
   (1) roots   (3) flowers
   (2) stems   (4) leaves

4. One of the main functions of the roots is to
   (1) produce food
   (2) produce new plants
   (3) absorb water
   (4) carry nutrients from the leaves to the stem

5. Tropisms are best described as
   (1) the way plants move from place to place
   (2) stimuli that cause plants to respond
   (3) the way plants make their own food
   (4) growth in response to stimuli

6. Stomata are tiny openings in the underside of leaves through which
   (1) carbon dioxide enters the plant
   (2) food enters the plant
   (3) chlorophyll enters the plant
   (4) sunlight enters the plant

7. The **products** of photosynthesis are
   (1) sugar and oxygen
   (2) carbon dioxide and water
   (3) sugar and carbon dioxide
   (4) carbon dioxide and oxygen

8. When sap moves from the leaves to the roots it flows through the
   (1) xylem   (3) palisades layer
   (2) phloem   (4) stomata

Questions 9 and 10 are based on the diagram below, which shows how a seedling grows after it is turned on its side.

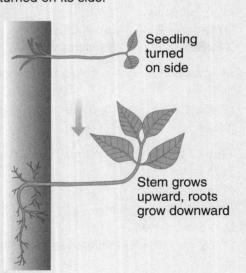

Seedling turned on side

Stem grows upward, roots grow downward

9. The stimulus that causes the stem to turn one way, and the roots to turn the opposite way, is
   (1) light   (3) gravity
   (2) touch   (4) water

10. Tropisms can be positive or negative. In this plant, the geotropism is
   (1) positive in both the roots and the stem
   (2) negative in both the roots and the stem
   (3) positive in the roots and negative in the stem
   (4) negative in the roots and positive in the stem

Thinking and Analzing (Part 2)

Base your answers to questions 1 and 2 on the paragraph below and on your knowledge of science.

Plants need water to carry out photosynthesis. Water is absorbed by the roots and brought to the leaves through the stems. In the leaves, water vapor can escape through the stomata. To prevent the loss of water, special cells called guard cells surround each stoma to open or close it. These guard cells open the stomata during the day and close it at night.

1. Why do the guard cells close the stomata at night?

2. Why must the stomata be open during the day?

3. The diagram below illustrates a geranium leaf that has been partially covered with black paper for three days.

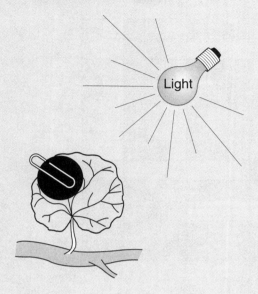

When the black paper is removed, the area that was covered by the paper has turned white. The white section of the leaf tests negative for the presence of sugar, and the green section tests positive for the presence of sugar. Explain why the white and green sections of the leaf have different sugar test results.

Base your answers to questions 4 and 5 on the diagrams below.

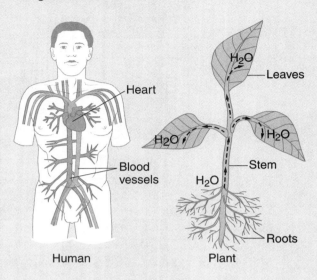

4. Select *one* structure labeled in the plant system above and explain how it contributes to the way the organism functions.

5. What plant tissues carry out the same functions in the plant as the blood vessels in the human?

6. Some plants have green stems, while other plants have woody stems. What process can occur in green stems, but *not* in woody stems?

7. Identify the plant organ that is responsible for
   (a) anchoring the plant
   (b) making food
   (c) absorbing water
   (d) letting in carbon dioxide
   (e) supporting the leaves
   (f) transporting food to the roots

Chapter Puzzle (*Hint:* The words in this puzzle are terms used in the chapter.)

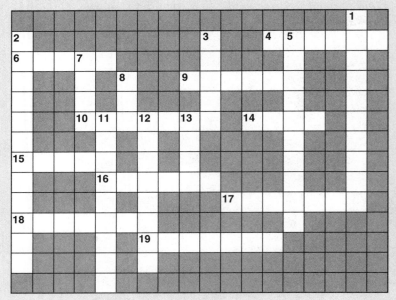

Across

4 the tissue that carries materials downward through a plant

6 ___tropism is the movement of a plant toward or away from water

9 the part of a plant that contains its reproductive organs

10 the growth of a plant toward or away from a stimulus

14 the plant organ where most photosynthesis takes place

15 ___tropism is the movement of a plant toward or away from light

16 ___tropism is the response of a plant to touch

17 plants that live less than one year

18 openings in leaves that permit the exchange of gases

19 inactive

Down

1 plants that live for several years

2 structures in plant cells that are the sites of photosynthesis

3 the tissue that carries materials upward through a plant

5 green stems on non-woody plants are called ___ stems

7 the plant organ that absorbs water and anchors the plant

8 ___tropism is the movement of a plant toward or away from the pull of gravity

11 the structures that absorb water into the root are called the ___ ___

12 the region within a leaf where most photosynthesis occurs is called the ___ layer

13 a plant organ that supports the plant and transports materials

Chapter 13

One-Celled Organisms

Contents

Flagellate bacteria.

What Is This Chapter About?

All of the organisms discussed in Chapters 10, 11, and 12 were multicellular. Multicellular organisms consist of many cells working together to keep the organism alive. Each organism is made of several different kinds of cells. Each type of cell—muscle cell, nerve cell, and blood cell—has a different structure. Can you imagine if an entire organism consisted of just one cell? That cell would have to carry out *all* of the life processes. It would need to eat, excrete, produce energy, reproduce, and react to changes in its environment. Welcome to the world of unicellular organisms.

In this chapter you will learn:

1. Unicellular organisms are found in the kingdoms monera, protista, and fungi.

2. Bacteria belong to the kingdom monera.

3. Bacteria are prokaryotes, cells that do not have a nucleus.

4. Protists include protozoa and algae.

5. Most protozoa have special structures that enable them to move from place to place.

6. Protozoa contain structures called vacuoles that aid in digestion and excretion.

7. Algae make their own food through photosynthesis.

Science in Everyday Life

Do your feet smell? The organisms that cause this odor are bacteria, and you are literally covered with them! Scientists estimate that at least 180 different kinds of bacteria live on your skin. Fortunately, most of these are harmless. Some bacteria, however, do cause serious illnesses.

Internet Sites:

http://www.microscopy-uk.org.uk/mag/ wimsmall/small.html Visit this Web site for a microscopic view of protozoa, algae, and bacteria. Click on the drop of water at the top of the page, and then examine each type of organism.

http://biology4kids.com/files/micro_main.html Go to each stop on the "Site Tour" to learn about many types of microorganisms.

13.1 What Are Unicellular Organisms?

Objectives

Distinguish unicellular organisms from multicellular organisms.

Identify unicellular organisms and the kingdoms to which they belong.

Distinguish between prokaryotes and eukaryotes.

Terms

monera (moh-NAIR-uh): a kingdom of one-celled organisms whose cells do not have a nucleus

prokaryotes (proh-KAH-ree-uhts): cells that do not contain a nucleus

protista (pro-TEES-tuh): a kingdom of simple plant-like and animal-like organisms whose cells contain a nucleus

eukaryotes (you-KAH-ree-uhts): cells that contain a nucleus

algae (AL-jee): plant-like protists that make their own food through photosynthesis

protozoa (proh-toh-ZOH-uh): one-celled, animal-like protists

Monera

Unicellular organisms are found in three different kingdoms. Bacteria (singular: bacterium) are the simplest unicellular organisms, and they belong to the kingdom **monera**. The kingdom monera contains organisms whose cells do not have a nucleus. In these cells, the chromosomes, which contain their genetic material, are found coiled within the cytoplasm. (See Figure 13.1-1.)

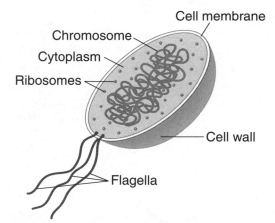

Figure 13.1-1. A typical bacterium.

Bacteria cells are **prokaryotes**. The word "prokaryote" means "before nucleus." Prokaryotes may be the oldest forms of living things. The name indicates that they existed before there were cells that contained a nucleus.

Figure 13.1-1 shows a typical bacteria cell. Notice that there is no nucleus. There are also no mitochondria and no vacuoles. Respiration and excretion occur throughout the cytoplasm.

Protista

The kingdom **protista** contains many different unicellular organisms. Some of these are more animal-like, while others are more plant-like. Figure 13.1-2 shows some types of protists. Notice that each of these cells contains a nucleus. For this reason, these cells are called **eukaryotes**. The prefix *eu-* means "true." These are cells that contain a "true" or clearly defined nucleus.

One of the protists in Figure 13.1-2, the euglena, contains chloroplasts. *Stop and Think:* What other cells have you studied that also contain chloroplasts? *Answer:* Plant cells contain chloroplasts. Like green plants, euglena use photosynthesis to make their own food. Protists that can make their own food through photosynthesis are called **algae**.

The ameba and the paramecium cannot make their own food. Protists that must take in food are **protozoa** (singular: *protozoan*). The name protozoa means "first animals" (*proto-* means "first" and *zoa* means "animals").

Fungi

Fungi is a kingdom of organisms whose cells contain cell walls but no chloroplasts. Plant cells also contain cell walls, but because fungi *cannot* perform photosynthesis, they are not considered plants.

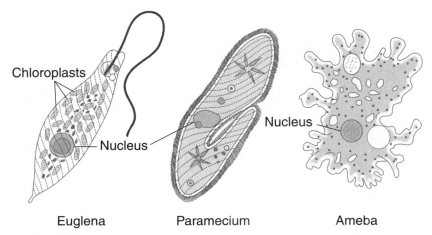

Figure 13.1-2. Some protists.

Most fungi, like molds and mushrooms, are multicellular. Yeasts, however, are unicellular organisms that contain cell walls but no chloroplasts. Yeasts are one-celled fungi. (See Figure 13.1-3.) *Stop and Think:* Are yeast cells prokaryotes or eukaryotes? *Answer:* Since yeast cells contain a nucleus, they are eukaryotes.

There is as much diversity in the tiny world of unicellular organisms as there is in multicellular organisms.

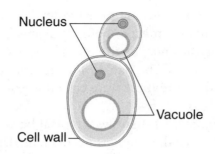

Figure 13.1-3. Two yeast cells.

Activity

A *dichotomous key* organizes information by asking a series of yes-no questions. It helps you to identify an organism by offering two choices at each step. Your choice leads you to the next question and, finally, to a correct conclusion.

The chart below uses dichotomous keys to classify organisms into one of the five kingdoms. Copy the chart into your notebook, and complete it by writing the name of the correct kingdom in each oval. One kingdom will be used twice.

Create your own dichotomous key to organize the kingdoms. In your chart, begin with the question, "Is it unicellular?" You will find that in your chart, a **different** kingdom appears twice! You might even need to use **two** kingdoms twice.

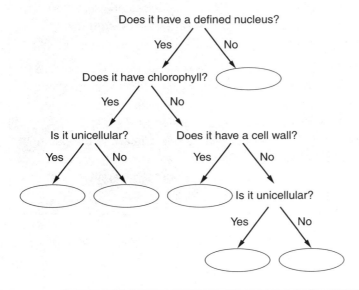

Interesting Facts About Classification —The Sixth Kingdom?

At one time, all living things were divided into just two kingdoms, plants and animals. In this system, bacteria and algae were considered plants, while protozoa were considered animals. By 1894, protists had been added, and then, by 1959, fungi and monera were added. Five kingdoms seemed to be enough.

Then, in 1977, scientists found a new kind of bacteria. These bacteria live under the most extreme conditions, such as very high temperatures, high acidity, and very salty water. Careful examination of these bacteria showed them to be very different from all other bacteria. Scientists then divided the bacteria into two groups—*eubacteria* (you-back-TEER-ee-uh) and *archaebacteria* (ahr-key-back-TEER-ee-uh). The bacteria that most people are familiar with are eubacteria. All disease-causing bacteria are eubacteria. (Remember that the prefix *eu-* means "true." The prefix *archae-* means "old.") Scientists believe that archaebacteria may be the earliest form of life. Biologists created a special sixth kingdom, the archaebacteria, because these organisms are so unique.

As scientists gather more information about living things, they continue to adjust the system of classification. Recently biologists have added a new level to the system, the domain. Domains are even larger than kingdoms. There are now three domains, archaea, bacteria, and eukarya. Archaebacteria are in the archaea domain. Eubacteria are in the bacteria domain. Protists, fungi, plants, and animals are all part of the eukarya domain. As biologists learn more and more about the diversity of living things, these new domains may change as well!

Questions

1. Which organisms are prokaryotes?
   (1) fungi   (3) algae
   (2) bacteria   (4) humans

2. Which organisms may be multicellular?
   (1) fungi   (3) ameba
   (2) bacteria   (4) paramecia

3. The illustration below shows four different kinds of organisms. Which of these organisms is multicellular?
   (1) hydra   (3) paramecium
   (2) euglena   (4) ameba

4. Algae and protozoa belong to the kingdom
   (1) monera   (3) protista
   (2) fungi   (4) animalia

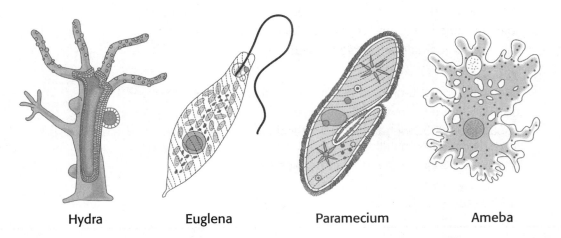

Hydra   Euglena   Paramecium   Ameba

Thinking and Analyzing

1. Explain the difference between prokaryotes and eukaryotes.

2. At one time, scientists classified all living things into just two kingdoms—plant and animal. Using this system, where would you place a euglena (see Figure 13.1-2)? Justify your decision.

13.2 How Do Unicellular Organisms Get Around?

Objectives

Describe three methods of locomotion in unicellular organisms.

Identify some stimuli and responses in unicellular organisms.

Compare three different types of bacteria.

Terms

flagellum (fluh-JEL-uhm): a tail-like structure used for locomotion

cilia (SIH-lee-uh): tiny hair-like structures

pseudopod (SOO-duh-pahd): an extension of the cytoplasm of an ameba, used in locomotion and nutrition

bacilli (bah-SIHL-eye): rod-shaped bacteria

cocci (COX-eye): spherical bacteria

spirilla (spy-RILL-uh): spiral-shaped bacteria

Look at the three organisms in Figure 13.1-2 on page 387. All of these organisms live in water, and each has its own method of moving from place to place. The structures that enable each organism to move are shown in the figure. Can you find these structures?

Flagellum

Figure 13.2-1 is a labeled diagram of an euglena. The euglena has a tail-like structure called a **flagellum** (plural: *flagella*). The flagellum spins like a tiny propeller, moving the euglena from place to place. Look at the bacterium in Figure 13.1-1 on page 386.

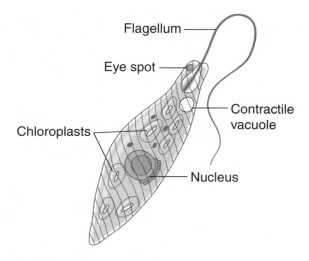

Figure 13.2-1. An euglena.

This particular bacterium has three flagella. Not all bacteria have flagella, and some cannot move at all. Organisms that are propelled by flagella are called *flagellates.*

Cilia

Look at the paramecium in Figure 13.2-2. Can you see how the paramecium moves around? Hair-like structures called **cilia** provide locomotion. The cell is surrounded by these cilia, which act like oars to propel the organism through its watery environment. Organisms that are propelled by cilia are called *ciliates.*

Pseudopods

Figure 13.2-3 is a labeled diagram of an ameba. Can you guess how the ameba moves? It has no cilia and no flagella. However, the body of the ameba is very flexible. It can change its shape to form an extension called a **pseudopod.** (*Pseudo* means "false" and *pod* means "foot." A pseudopod is a "false foot.") The pseudopods move in one direction, and the rest of the cell follows. As the ameba moves, new pseudopods form, and the old pseudopods disappear back into the cell.

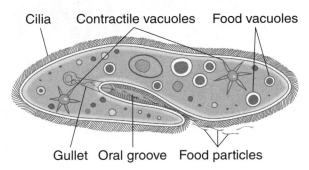

Cilia   Contractile vacuoles   Food vacuoles

Gullet   Oral groove   Food particles

Figure 13.2-2. A paramecium.

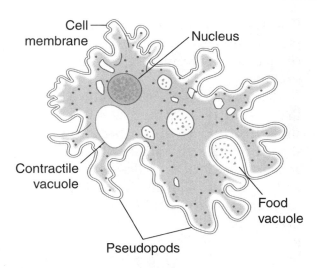

Cell membrane — Nucleus

Contractile vacuole

Food vacuole

Pseudopods

Figure 13.2-3. An ameba.

Responding to Stimuli

The movement of unicellular organisms helps them maintain homeostasis. They can move *toward* food and *away* from wastes. Some of these organisms can also respond to other changes in their environment. For example, euglena move *toward* a source of light. ***Stop and Think:*** Why does moving toward a light source help the euglena maintain homeostasis? ***Answer:*** Euglena use light to make their own food through photosynthesis. Euglena have a special structure called an eyespot, which can detect light.

Paramecia eat bacteria. Some of these bacteria produce acidic waste products. Paramecia move toward solutions that are slightly acidic. To a paramecium, "acid" means "food."

Bacteria

Even the simplest organisms, the bacteria, can respond to their environment. You have learned that some bacteria are flagellates. They can move from place to place in response to food, light, chemicals, temperature, or even electric currents.

Bacteria are classified according to their shape. (See Figure 13.2-4.) Rod-shaped bacteria are called **bacilli**, spiral-shaped bacteria are called **spirilla**, and spherical bacteria are called **cocci**. Bacilli and spirilla are sometimes flagellates, but cocci are not. These spherical bacteria have no structures that provide locomotion. One particular type of bacteria, *streptococci*, causes the disease strep throat. ***Stop and Think:*** If streptococci have no method of locomotion, how do they move through your body? ***Answer:*** These bacteria cannot move by themselves, but your body can move them. Once the bacteria get into your bloodstream, they can "go with the flow." As your blood brings food and oxygen to your cells, it may also bring harmful bacteria.

Good and Bad Bacteria

Not all bacteria cause disease. Most bacteria are harmless, and some are even beneficial. Of course, the bacteria do not know whether they are helping us or harming us. They are just trying to find food and maintain homeostasis.

One very common type of bacteria is the bacillus known as *E. coli*. These bacteria are found in the human digestive tract, where they provide vitamins and aid in digestion. Other bacteria found in the soil break down the remains of dead plants and animals. These *bacteria of decay* return nutrients to the soil to be reused by living plants. Still other bacteria are used in the manufacture of yogurt, cheese, and sourdough bread.

On the other hand, some bacteria do cause a number of serious diseases. Strep throat, tuberculosis, Lyme disease, and pneumonia are all caused by bacteria. Bacteria also cause many types of food poisoning, including salmonella and botulism.

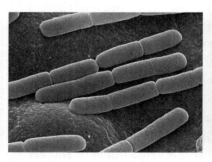  

(a) Bacilli   (b) Cocci   (c) Spirilla

Figure 13.2-4. Three types of bacteria.

Activity

Visit the Web site
http://www.teachersdomain.org/sci/life/stru/cellstruct/index.html
on the Internet. Watch the *Single Cell Organism* video by
clicking on the View button. Then click "Take a test drive," and
click the View button again. (If you get a window that says
"MIME type configuration" click NO.) You can Turn Captions
ON to get help with the spelling of some new words.
Use the information in the video to answer the following
questions:

1. What are three things that are accomplished by the cilia?
2. How does an ameba move?
3. What organelle controls water regulation in the paramecium?
4. What organelle in an euglena helps it to find sunlight?

Interesting Facts About Malaria

What animal kills the greatest number of people each year? Most likely, it is the mosquito! Mosquito bites result in diseases that kill as many as two million people per year. The majority of these deaths are due to malaria, which infects between 350 and 500 million and kills more than 1 million people per year. In Africa, malaria is the leading cause of death in children under the age of 5.

Why does a discussion of malaria belong in this chapter? The mosquito does not cause malaria; it just spreads the disease from one person to another. Malaria is actually caused by a protozoan called *plasmodium* (singular). When the plasmodium gets into the bloodstream, it infects the red blood cells and the liver. It reproduces, sending more plasmodia (plural) into the bloodstream. After a mosquito bites an infected person, it carries the protozoan to everyone else it bites.

Malaria is spread by only one species of mosquito. If we could eliminate that mosquito, or protect people from its bite, malaria would disappear. The United Nations has declared the decade 2001 to 2010 "The Decade to Roll Back Malaria." The lives of millions of African children depend on the success of this campaign. To learn more about the fight against malaria, visit the Web site
http://www.malarianomore.org/kids/index.php

Questions

1. The "false feet" of the ameba are its
   (1) cilia   (3) flagella
   (2) pseudopods   (4) bacilli

2. Euglena and some bacteria move using whip-like structures called
   (1) cilia   (3) flagella
   (2) pseudopods   (4) bacilli

3. The round-shaped bacteria that cause strep throat are classified as
   (1) cocci   (3) bacilli
   (2) cilia   (4) spirilla

4. Which organism is a ciliate?
   (1) bacteria   (3) paramecia
   (2) ameba   (4) euglena

5. Which statement about bacteria is most accurate?
   (1) All bacteria cause disease.
   (2) All bacteria are flagellates.
   (3) All bacteria are helpful to humans.
   (4) All bacteria are unicellular.

Thinking and Analyzing

1. What two structures enable an euglena to move toward a source of light?

2. Complete the analogy: Cilia are to oars, as ___ are to propellers.

3. Describe how an ameba moves.

13.3 How Do Unicelluar Organisms Get Nutrients and Elimate Wastes?

Objectives

Explain nutrition in unicellular organisms.

Explain excretion in unicellular organisms.

Terms

food vacuole (VACK-yoo-ohl): an organelle in which digestion takes place

phagocytosis (FAG-oh-sy-TOH-sis): the process by which a cell takes in food by surrounding it and enclosing it in a food vacuole

contractile (kuhn-TRAK-tile) vacuole: an organelle found in protozoa that removes water from the cell

---

Nutrition and excretion are life processes common to all living things. The single cell in a unicellular organism must be able to find, take in, and digest food. It must also be able to get rid of its wastes. In this lesson, we look at the same organisms we examined in the last lesson, but this time we see how they eat and excrete.

Nutrition

Paramecium

If you look back at Figure 13.2-2 on page 393, you will see how paramecia get their food. The *oral groove* serves the same purpose as a mouth. The cilia sweep food, usually bacteria, into the oral groove. The food passes through the *gullet*, and a circular structure called a **food vacuole** forms at the end of the gullet. The food vacuole contains enzymes that digest the food. Nutrients then pass from the food vacuole into the cytoplasm, where the organism uses them.

Ameba

Amebas eat bacteria, algae, and other protozoa. Look at the ameba in Figure 13.3-1. You can see two pseudopods around the food particle. The pseudopods surround the food and form a food vacuole, in a process called **phagocytosis**. *Lysosomes* are organelles that contain the enzymes needed to digest the food. Lysosomes move to the food vacuole, where they secrete their enzymes. Digestion occurs inside the food vacuole. To see a photograph of an ameba

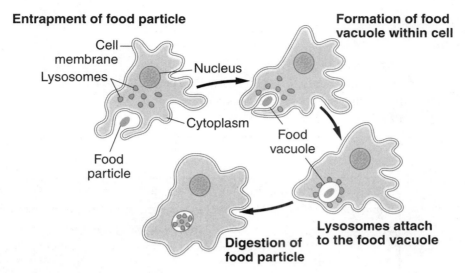

Figure 13.3-1. Phagocytosis in an ameba.

capturing its prey, visit *http://www. microscopy-uk.org.uk/mag/artsep01/ feed.html* on the Internet.

Euglena

Recall that euglena have chloroplasts. If there is enough light, euglena do not need to "eat" at all! They can produce their own food through photosynthesis. Euglena can also take in food through the process of phagocytosis if light is not available. The cell membrane of the euglena can surround smaller organisms and form a vacuole in the cytoplasm.

Bacteria

Some bacteria are able to make their own food. Most, however, need to "eat." Bacteria do not have a mouth or pseudopods to help them collect food. Nutrients pass directly into the organism through its cell membrane. Before this can happen, the food must already be broken down. Bacteria

actually digest their food before they eat it! When bacteria come in contact with food, they secrete enzymes that pass out of their cells and into the food. After the food has been digested, the bacteria can absorb the nutrients.

Excretion

Excretion in the paramecium, ameba, and euglena is very similar. All of these organisms live in freshwater, which is constantly diffusing through their cell membranes and into their cells. In order to control the amount of water in their cells, these protists have organelles called **contractile vacuoles**. These vacuoles look like large bubbles inside the cytoplasm. The contractile vacuole collects and removes water. It fills with fluid and then moves to the cell membrane, the skin of the cell. At the membrane, the vacuole contracts, and squeezes out its excess water. Other wastes also diffuse out of the organism through the cell membrane.

Activity

Visit the Web site
http://www.childrensuniversity.manchester.ac.uk/interactives/
science/microorganisms/whatareprotozoa.swf (*Note:* this is a
British site. Some of the words and spelling may seem strange.)
Take the quiz at the end of the animation.

Interesting Facts About Excretion in Bacteria

Bacteria remove wastes the same way they take in food—directly through their cell membrane. *What* bacteria excrete may be more interesting than *how* bacteria excrete. For example, *micrococci* are bacteria that live on your skin. Like many types of bacteria, they thrive in a moist, dark environment, such as the area between your toes. They get their nutrients from your sweat. These bacteria excrete the foul odor that we call "smelly feet." The more you sweat, the more the bacteria eat, grow, and excrete. As the amount of waste increases, the odor gets worse. On other parts of your body, sweat evaporates, but your shoes and socks trap the sweat, providing a feast for the bacteria.

Questions

1. Which statement best describes the role of the pseudopods in the ameba?
   (1) They are used only for locomotion.
   (2) They are used only to obtain food.
   (3) They are used both for locomotion and to obtain food.
   (4) They are used only for excretion.

2. Which organism has a mouth-like structure that takes in food?
   (1) paramecium   (3) euglena
   (2) ameba   (4) bacterium

3. The main function of the contractile vacuole is to
   (1) take in water
   (2) take in food
   (3) remove water
   (4) digest food

4. Which organism digests its food *outside* its body?
   (1) paramecium   (3) euglena
   (2) ameba   (4) bacterium

5. Digestion in the ameba takes place in the
   (1) nucleus
   (2) cell membrane
   (3) food vacuole
   (4) contractile vacuole

Thinking and Analyzing

Base your answers to questions 1 and 2 on the reading passage below and on your knowledge of science.

> Many protozoa live in freshwater. The surrounding water moves into their cells through a process called osmosis (*oz-MOH-sis*). *Osmosis* is the movement of water through a membrane. Osmosis occurs constantly. With water constantly moving into their cells, freshwater protozoa must constantly remove water from their cells. Otherwise, the cells would burst like an overfilled balloon.

1. How does osmosis affect the amount of water in the cells of freshwater protozoa?

2. How do freshwater protozoa get rid of excess water?

3. Compare the way an ameba and a paramecium take in food.

Review Questions

Term Identification

Each question below shows two terms from Chapter 13. One of the terms is defined.
(1) Choose the term that matches the definition.
(2) Describe how the two terms are different.
Following each term is the section (in parentheses) where the description or definition of that term is found.

1. *Monera (13.1) — Protista (13.1)*
   A kingdom of one-celled organisms whose cells do not have a nucleus

2. *Eukaryotes (13.1) — Prokaryotes (13.1)*
   Cells that contain a nucleus

3. *Algae (13.1) — Protozoa (13.1)*
   One-celled, animal-like protists

4. *Flagella (13.2) — Cilia (13.2)*
   Tiny hair-like structures

5. *Bacilli (13.2) — Cocci (13.2)*
   Rod-shaped bacteria

6. *Pseudopod (13.2) — Phagocytosis (13.3)*
   The process by which a cell takes in food by surrounding it and enclosing it in a food vacuole

7. *Contractile vacuole (13.3) — Food vacuole (13.3)*
   An organelle found in protozoa that removes water from the cell

Multiple Choice (Part 1)

Choose the response that best completes the sentence or answers the question.

1. Which term describes bacteria, but *not* protozoa?
   (1) unicellular   (3) prokaryote
   (2) multicellular   (4) eukaryote

2. Flagella and cilia are both structures used in
   (1) respiration   (3) nutrition
   (2) excretion   (4) locomotion

3. Which term includes the other three?
   (1) protista   (3) algae
   (2) protozoa   (4) ameba

4. Which organism is most likely to move *toward* a source of light?
   (1) ameba   (3) paramecium
   (2) euglena   (4) yeast

5. Which one-celled organism is part of the fungi kingdom?
   (1) ameba   (3) paramecium
   (2) euglena   (4) yeast

6. Which cell structure identifies a cell as a eukaryote?
   (1) nucleus   (3) cell membrane
   (2) mitochondria   (4) chloroplast

7. An organism that is described as a flagellate is most likely to have
   (1) one or more tail-like structures
   (2) pseudopods
   (3) hundreds of hair-like structures
   (4) a spiral shape

8. Which type of bacteria is correctly matched with its shape?
   (1) bacilli—round
   (2) spirilla—rod-shaped
   (3) cocci—spiral
   (4) cocci—spherical

9. Excretion of water in a paramecium is controlled by the
   (1) cilia
   (2) food vacuole
   (3) contractile vacuole
   (4) oral groove

10. Why did the teacher stop telling jokes about paramecia?
   (1) She couldn't find the groove.
   (2) The students protist-ed.
   (3) The jokes got cilia and cilia.
   (4) All of these.

Thinking and Analyzing (Part 2)

1. Identify *three* kingdoms that contain unicellular organisms. Identify *one* unicellular organism from each kingdom.

2. Identify *one* type of protist that is constantly changing its shape. Why does it do this?

3. Euglena and paramecia are both organisms that live in freshwater. Compare these two organisms by identifying *two* similarities and *two* differences between them.

4. Bacteria are both harmful and helpful to humans. Identify *one* way bacteria can be helpful, and *one* way they can be harmful.

5. *Streptococcus* is the name of the bacterium that causes strep throat. *Bacillus anthracis* is the bacterium that causes a serious disease called anthrax. *Campylobacter* is a bacterium that causes food poisoning. These three bacteria illustrate the three common shapes of bacteria. Based on the names of the first two, describe the shapes of all *three*.

6. What are *two* important functions of the cilia of the paramecium?

7. Briefly describe the functions of the food vacuole and the contractile vacuole in freshwater protozoans.

Chapter Puzzle (*Hint:* The words in this puzzle are terms used in the chapter.)

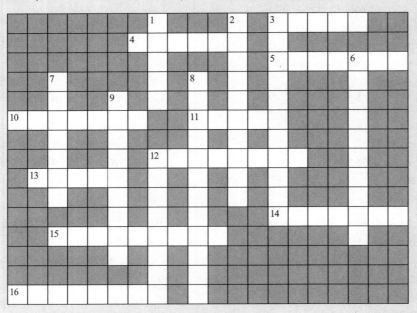

Across

3 tiny hair-like structures

4 a kingdom of organisms whose cells do not have a nucleus

5 prokaryotes are cells that do not contain a ___

10 rod-shaped bacteria

11 a one-celled organism that uses pseudopods to trap its prey

12 the kingdom that contains protozoa and algae

13 plant-like protists that make their own food through photosynthesis

14 a flagellate protist that contains chloroplasts

15 false foot

16 spiral-shaped bacteria

Down

1 spherical bacteria

2 a ciliate protozoan that has an oral groove

3 the vacuole that controls water in the cell is called the ___ vacuole

6 cells that contain a nucleus

7 a food ___ is where digestion takes place in a cell

8 the process by which a cell surrounds food and makes a vacuole

9 a tail-like structure used for locomotion

12 one-celled, animal-like protists

Glossary

A

absorption: in physics, the process by which light rays are absorbed and transformed into heat; in biology, the movement of material into the blood

algae (AL-jee): plant-like protists that make their own food through photosynthesis

alveoli (al-vee-OH-lie): tiny air sacs in the lungs

amplitude: the height of a wave measured vertically from the undisturbed surface to the top of the wave

annuals: plants that live less than one year

arteries (AHR-tuh-rhees): blood vessels that carry blood away from the heart

arthropod: an invertebrate with legs that have joints

ash: fine dust particles produced by a volcanic eruption

atom: the smallest piece of an element that has the properties of that element

atomic number: the number of protons in an atom

atria (AY-tree-uh): upper chambers of the heart that receive blood

B

bacilli (bah-SIHL-eye): rod-shaped bacteria

bedrock: the solid rock portion of the crust

biodiversity (buy-oh-di-VUR-si-tee): variety in living things

biosphere (BY-oh-sfeer): zone of living things that surrounds Earth

bone: hard, rigid connective tissue that supports and protects the organs of the body

breathing: moving air into and out of the lungs

brittle: breaks when hit with a hammer

burning: a process in which a material reacts quickly with oxygen to produce energy

C

caldera (kal-DER-uh): a basin formed by the collapse of a volcanic mountain

capillaries (CAP-uh-lehr-ees): tiny blood vessels that connect arteries to veins

cardiac muscle: involuntary muscle found only in the heart

cartilage (KAR-tuh-lij): flexible connective tissue that cushions bones

cast: a mold filled with new mineral matter

cell: the building block of all living things

cell membrane: the outer covering of the cell that controls the flow of materials into and out of the cell

cell wall: the rigid outer covering of a plant cell

cellular respiration: the chemical reaction in the cells that combines glucose and oxygen to produce energy and form carbon dioxide

Cenozoic (sen-uh-ZOH-ik) **era:** period of time on Earth (present–65 million years ago) when recent life forms existed; mammals dominated

central nervous system: the brain and spinal cord

centrifuge (SEN-trih-fyooj): a machine that separates the parts of a mixture by spinning it

chemical bond: an attraction that holds two atoms together

chemical change: when matter changes to produce a new substance

chemical digestion: the process of breaking down complex substances in food into simpler substances that can be used by cells

chemical equation: a chemical reaction expressed in formulas and symbols

chemical formula: a way of representing a substance using symbols and numbers

chemical property: a property that involves the formation of a new substance

chitin (KY-tin): the hard material that makes up the exoskeleton

chlorophyll: a green chemical plants need for photosynthesis

chloroplasts: structures in plant cells where photosynthesis occurs

cilia (SIH-lee-uh): tiny hair-like structures

cinders: coarse rock particles produced by a volcanic eruption

cocci (COX-eye): spherical bacteria

coefficient (koh-uh-FISH-uhnt): a number placed before a formula in a balanced chemical equation

composition: the material that something is made of

compound: a substance containing two or more different elements

concentrated (KAHN-sehn-tray-ted): a strong solution

concentration (KAHN-sehn-TRAY-shun): the strength of a solution

connective tissue: tissue that supports the body and holds it together

conserved (kuhn-SURVED): kept the same

contour interval: elevation difference between two adjacent contour lines

contour (KAHN-toor) **lines:** lines connecting points of equal elevation

contractile (kuhn-TRAK-tile) **vacuole:** an organelle found in protozoa that removes water from the cell

convection cell: the motion of Earth's mantle caused by the upward heat transfer inside Earth

convergent (kahn-VER-gent) **boundary:** the crustal plate boundary produced when crustal plates come together

core: the center region of Earth

covalent (koh-VAY-luhnt) **bond:** a chemical bond in which electrons are shared

crater: a depression at the top of a volcanic cone

crop: an organ in the digestive tract that stores food

crust: Earth's thin outermost layer that is composed of solid rock

crystal: mineral with a regular atomic structure and geometric shape

cytoplasm (SYE-toe-plaz-uhm): the gel-like substance that fills the cell

D

decibel: unit used to measure the intensity or loudness of sound

diaphragm (DIE–uh-frahm): a muscle in the chest that controls breathing

diffusion (dih-FYOO-zhin): the movement of material from an area of high concentration to an area of low concentration

digestion: the process of breaking down food into a form that can be used by the cells

digestive enzymes (EHN-zimes): chemicals that aid in chemical digestion

dilute (die-LOOT): a weak solution

dissolve: to go into solution

distillation (diss-tehl-LAY-shun): a method that separates the parts of a mixture by evaporation and condensation

divergent (di-VER-gent) **boundary:** the crustal plate boundary produced when crustal plates move apart

Doppler (DAH-pluhr) **effect:** a change in wave frequency caused by the relative distance between the sound source and the listener

dormant: inactive

ductile (DUCK-tile): can be stretched into wires

E

earthquake: a sudden shaking in Earth's crust

electromagnetic (ih-lehk-troh-mag-NEH-tihk) **energy:** different types of radiation that move in waves and can travel through space

electromagnetic spectrum: a continuous band of electromagnetic waves arranged according to wavelength and frequency

electron: a negatively charged subatomic particle found outside the nucleus

element: a substance that cannot be broken down into a simpler substance

elimination: the removal of solid wastes from the body

endocrine (EHN-doh-krin) **gland:** an organ that secretes hormones

endoskeleton (en-doh-SKEL-ih-tin): the system of bones that is found inside an animal

epicenter (EHP-ih-sehnt-uhr): the point on Earth's surface directly above the focus of an earthquake

erosion (ih-ROH-zhuhn): the process that occurs when particles of rock are removed from the rock and carried away

eukaryotes (you-KAH-ree-uhts): cells that contain a nucleus

event: change in one sphere that affects another sphere

excretion: the process of removing metabolic wastes from the body

exhale: to breathe out

exoskeleton (ek-soh-SKEL-ih-tin): the hard outer covering found on arthropods

F

feces (FEE-seez): solid wastes

filtration (fihl-TRAY-shun): a method that separates the parts of a mixture based on particle size

flagellum (fluh-JEL-uhm): a tail-like structure used for locomotion

flammability (flah-muh-BILL-ih-tee): the ability to burn

flower: the part of the plant that contains its reproductive organs

focus (FOH-kuhs): the underground location where an earthquake occurs

food vacuole (VACK-yoo-ohl): an organelle in which digestion takes place

fossils: the remains or traces of organisms that lived long ago

frequency: the number of waves that pass by a fixed point in a given amount of time

fuel (FYOOL): a substance that can be used to produce energy

G

geologic time scale: a detailed chart that shows the geologic and biological history of Earth

geotropism (JEE-oh-TROH-piz-uhm): the growth of a plant toward or away from the pull of gravity

gills: structures in fish that remove dissolved oxygen from water

gizzard: an organ in the digestive tract that grinds food

group: a column in the periodic table of the elements

H

half-life: the amount of time it takes for half the atoms of a radioactive substance to change into nonradioactive atoms

hemoglobin (HEE-muh-gloh-bin): a protein in red blood cells that carries oxygen

herbaceous (huhr-BAY-shuhs) **stem:** green stems found on non-woody plants

hertz (HURTS): unit used to measure the frequency of waves

homeostasis (hoe-mee-oh-STAY-sis): maintaining a constant internal environment

hormone (HAWR-mone): a substance secreted by an endocrine gland that has an effect on the body

hydrotropism (HIGH-droh-TROH-piz-uhm): the growth of a plant toward or away from water

hyperthermia (HY-puhr-THUR-mee-uh): a higher-than-normal body temperature

hypothermia (HY-poh-THUR-mee-uh): a lower-than-normal body temperature

I

igneous (IHG-nee-uhs) **rocks:** rocks that are produced by the cooling and hardening of hot, liquid rock

index fossil: the fossil of an animal or plant that existed for a relatively short time and lived over a wide area, used to match and date rock layers

infection: the illness or damage to cells that is caused by bacteria, viruses, or other microscopic organisms

inhale: to breathe in

inorganic: a substance that is not living and never was living

insoluble: *not* able to dissolve in a solvent

instincts (IN-stinkts): complex behaviors that are automatic and unlearned

intensity: the amount of energy in a sound wave, that is, how loud it sounds, which is determined by the amplitude of the sound wave

invertebrate (in-VER-tuh-brit): an animal that does not have a backbone

involuntary muscles: muscles that are *not* controlled by your conscious mind

ion (EYE-on): a positively or negatively charged particle formed by the loss or gain of electrons

ionic (eye-ON-ick) **bond:** a chemical bond formed by the attraction of oppositely charged ions

L

landform: a large land feature defined on the basis of how high it is, the steepness of its slope, and the type of rock it is made of; examples of common landforms are mountains, plains, and plateaus

Law of Conservation of Mass: the total mass of all the reactants is equal to the total mass of all the products

Law of Conservation of Matter: matter can be neither created nor destroyed during a chemical reaction

leaf: the plant organ where most photosynthesis takes place

lens: a piece of transparent glass or plastic that has a curved surface

life functions: processes that occur in all living things

ligament: connective tissue that connects bones to other bones

light: a form of radiant energy we can detect with our eyes

longitudinal (lahn-juh-TOO-duhn-uhl) **wave:** a wave that vibrates back and forth (push and pull) in the same direction as its direction of travel

luster: how a mineral looks when it reflects light; shininess

lymph (LIMF): a fluid that surrounds the body cells

M

malleable (MAL-ee-uh-bull): able to be hammered into shapes

Malpighian (mal-PIHG-ee-uhn) **tubules:** excretory organs found in insects

mammal: a warm-blooded animal that has hair covering its body and provides milk for its young

mantle: a layer of Earth that lies between the crust and the core

matter: anything that has mass and takes up space

mechanical digestion: the physical breakdown of food into smaller pieces

medium: a substance, such as air or water, through which waves travel

metabolic (met-uh-BOL-ik) **wastes:** harmful products produced by chemical reactions in cells

metabolism (muh-TAB-uh-liz-uhm): the sum of all chemical reactions that occur in an organism

metals: shiny, malleable elements that conduct electricity; metals are found on the left side of the periodic table

metamorphic (meht-uh-MAWR-fihk) **rocks:** rocks that are produced when pre-existing rocks undergo a change in form or composition because heat and pressure are applied to them underground

Mesozoic (mez-uh-ZOH-ik) **era:** period of time on Earth (65–251 million years ago) when "middle life" forms existed; reptiles (dinosaurs) dominated

mid-ocean ridge: a long chain of mountains on the ocean floor

mineral: a naturally occurring solid substance made of inorganic material with a definite chemical composition (chemical formula)

mixture: the result you get when you put two or more different substances together and they do not form a new substance

mold: a hollow area in sedimentary rock produced where a fossil dissolved out of the rock; an imprint of a fossil

molecule: a particle that contains two or more atoms bonded together

monera (moh-NAIR-uh): a kingdom of one-celled organisms whose cells do not have a nucleus

mountain: a landform that is significantly higher than the land surrounding it

multicellular (MUL-tea-CELL-yuh-ler): containing many cells

muscles: masses of tissue that contract to move bones and organs

N

nephridia (neff-RID-ee-uh): excretory organs found in earthworms

neurons (NUH-rahnz): nerve cells

neutron (NEW-tron): a neutral subatomic particle found in the nucleus

noble gas: elements found in group 18 of the periodic table that are nonreactive

nonmetals: elements found on the right side of the periodic table that are poor conductors of electricity

nucleus (NEW-clee-uhs; plural: *nuclei*, NEW-clee-eye): the structure within the cell that controls cell activities; the center of an atom containing the neutrons and protons

nutrient (NEW-tree-ehnt): a useful substance found in food

O

ore: mineral that is mined because it contains metals or other substances of value

organ: a group of tissues working together

organ system: a group of organs working together to carry out a specific life function

organelle (or-guh-NEHL): a small structure within the cytoplasm that carries out a specific life function

organic: a substance that is living now or was alive in the past

organism: a living thing

P

Paleozoic (pay-lee-uh-ZOH-ik) **era:** period of time on Earth (251–544 million years ago)

when early life forms existed; animals without backbones, fish, and amphibians dominated

palisade layer: the region within a leaf where most photosynthesis occurs

perennials (puhr-EN-ee-uhlz): plants that live for several years

perspiration: the fluid produced by sweat glands in the skin

phagocytosis (FAG-oh-sy-TOH-sis): the process by which a cell takes in food by surrounding it and enclosing it in a food vacuole

pheromones (FEHR-uh-mownz): chemicals released by an organism that allow it to communicate with other members of its species

phloem (FLOH-uhm): the tissue that carries materials downward through a plant

photosynthesis (foe-toe-SIN-thuh-sis): the chemical process by which green plants use sunlight to convert carbon dioxide and water into glucose and oxygen

phototropism: the growth of a plant toward or away from light

physical change: when a substance changes but does not form a new substance

physical property: a property of a substance that can be observed (such as color) and measured and that does **not** involve the formation of a new substance

pitch: the level of frequency of a sound, that is, how high or low it sounds, which is determined by the frequency of the sound wave

plain: a large, generally flat landform at a low elevation

plasma (PLAZ-muh): the liquid part of the blood

plateau (pla-TOE): a large, generally flat landform at a high elevation

platelet (PLAYT-lit): a blood cell that causes blood to clot

Precambrian (pree-KAM-bree-un) **era:** period of time on Earth (first 4 billion years) for which the fossil and rock records are poorly preserved

proboscis (proh-BAHS-kus): a hollow, tube-like organ used for sucking fluids

products: the materials you end up with after a chemical reaction

profile: a side view, or cross section, of a cutaway section of Earth's surface

prokaryotes (proh-KAH-ree-uhts): cells that do not contain a nucleus

property (PRAHP-uhr-tee): a characteristic used to describe matter

protista (pro-TEES-tuh): a kingdom of simple plant-like and animal-like organisms whose cells contain a nucleus

proton: a positively charged subatomic particle found in the nucleus

protozoa (proh-toh-ZOH-uh): one-celled, animal-like protists

pseudopod (SUE-duh-pahd): an extension of the cytoplasm of an ameba, used in locomotion and nutrition

R

radioactive dating: a method to determine the age of a rock by measuring its radioactivity

rate: the speed at which something happens

reactants: the materials that change in a chemical reaction

reflection: the bouncing of light rays off a surface

reflex (RHEE-flex): an automatic response designed to protect the body

refraction: the bending of light rays when they pass from one medium to another

relative age: the order or sequence in which rocks formed

respiration (rehs-puh-RAY-shun): a chemical process in which living things use glucose and oxygen to produce carbon dioxide, water, and energy.

rock: naturally occurring material composed of one or more minerals

rock cycle: the never-ending series of natural processes that change one type of rock into another type of rock

root: the plant organ that anchors the plant and absorbs water

root hairs: hair-like structures that absorb water into the root

S

saturated (SACH-uh-ray-ted): a solution that cannot dissolve any more solute at a given temperature

seafloor spreading: the way the seafloor moves, like a conveyor belt, away from a mid-ocean ridge

secretion: the process of producing chemicals and releasing them into the body

sedimentary (sehd-uh-MEHN-tuhr-ee) **rocks:** rocks that form from particles called *sediments* that pile up in layers, usually underwater

seismic (SIZE-mihk) **waves:** vibrations set off by an earthquake that travel through the earth

sense organs: organs that receive information about the environment

smooth muscle: involuntary muscles involved in digestion, respiration, and other body functions

soil: a mixture of small rock fragments, organic matter, water, and air

solubility (SAHL-you-BILL-uh-tee): the amount of solute that can dissolve in a given solvent at a given temperature

soluble (SAHL-yuh-buhl): able to dissolve in a solvent

solute (SAHL-yoot): in a solution, the substance that dissolves in the *solvent*; the solute may change its phase

solution: a *mixture* in which the particles of one substance are evenly distributed throughout a second substance

solvent (SAHL-vent): in a solution, the substance that dissolves the *solute*; the solvent always keeps its phase

sound: a form of mechanical energy produced by a vibrating object

spiracles (SPIR-ih-cullz): tiny openings in the bodies of insects that take in air

spirilla (spy-RILL-uh): spiral-shaped bacteria

stem: the plant organ that supports the plant and transports materials

stimulus: an environmental change that causes a response

stomata: openings in leaves that permit the exchange of gases

structural formula: a model that shows the arrangement of atoms in a molecule

subatomic particle: a particle that is smaller than an atom

subscript: a number, in a formula, written below and to the right of the symbol

substance: a pure form of matter with one set of properties

surface area: a measure of the amount of a substance that is exposed to other substances

suspension (suh-SPEN-shun): a mixture that separates if allowed to stand over a period of time

synapse (SIH-naps): the gap between one neuron and the next

T

tectonic forces: forces produced usually along plate boundaries that cause rocks to bend, crack, and uplift

tendon: cord-like tissue that connects muscles to bones

theory of continental drift: a theory that suggests that the continents were once a single mass of land that broke apart and are currently drifting across Earth's surface

theory of plate tectonics (tehk-TAHN-ihks): the theory that proposes that Earth's crust consists of moving plates

thigmotropism (THIG-moh-TROH-piz-uhm): the response of a plant to touch

tissue: a group of similar cells acting together to carry out a life process

topographic (tah-puh-GRAF-ihk) **map:** a map that shows the form and shape of the physical land features

transform boundary: the crustal plate boundary produced when crustal plates slide past each other

transmission: the passing of light rays through an object

transpiration (tranz-puh-RAY-shun): the escape of water molecules from the leaves of a plant

transverse wave: a wave that vibrates at a right angle (up and down) to the direction in which the wave is traveling

trench: a very deep, elongated gorge with steep walls in the ocean floor

tropism (TROH-piz-uhm): the growth of a plant toward or away from a stimulus

tsunami (tsoo-NAH-mee): a series of large ocean waves caused by an earthquake on the ocean floor

U

unicellular (YOU-nih-CELL-yuh-ler): containing only a single cell

unsaturated (UHN-sach-uh-ray-ted): a solution that can dissolve more solute at a given temperature

V

valence (VAY-lentz) **electrons:** electrons found in the outermost shell of an atom

valence shell: the outermost shell of an atom

vapor: the gas phase of a substance that is not normally a gas at room temperature

vascular (VASS-cue-luhr) **bundles:** the tissue in herbaceous stems that contains the xylem and phloem

veins: blood vessels that carry blood toward the heart

ventricles (VEHN-trih-kuhlz): lower chambers of the heart that pump the blood

vertebrae (VER-tuh-bray): bones that surround and protect the spinal cord

vertebrate (VER-tuh-brit): an animal that has a backbone

visible spectrum: series of colors that make up white light: red, orange, yellow, green, blue, and violet

volcanic (vahl-KAN-ic) **activity:** processes related to the movement of liquid rock underground (intrusive) and on Earth's surface (extrusive)

volcanic cone: a volcanic mountain

volcano (vahl-KAY-noh): an opening in Earth's surface where liquid rock is released

voluntary muscles: muscles that are controlled by your conscious mind

W

warm-blooded: an organism that maintains a constant body temperature

wavelength: the distance from the top of one wave to the top of the next wave

weathering: the breaking down of rocks into smaller pieces

word equation: a summary of a chemical reaction using words

X

xylem (ZY-luhm): the tissue that carries materials upward through a plant

Index

Photo Credits